THE TAO

OF SYSTEMS

THINKING:

Exploring the Parallels between

Eastern Mysticism and Systems Thinking

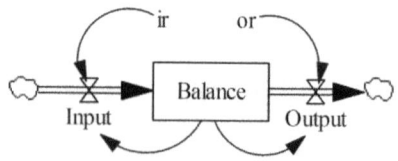

By Michael McCurley

The Tao of Systems Thinking:
Exploring the Parallels between Eastern Mysticism and
Systems Thinking

Copyright © 2015 by Michael McCurley

Printed by Create Space of Amazon.com

Three to the Fourth Power Publications

ISBN-13:978-1517128166
ISBN-10:1517128161

First Published/Printed in 2015, 2016.

Introduction to The Tao of Systems Thinking

This book explores the parallels between the wisdoms of ancient Chinese mysticism which were written in *Tao Te Ching* by *Lao Tzu more than 2,500 years ago, and developments in Systems Thinking in the mid-20th century with the advent of modern computers, through the work of *Jay W. Forrester and those he guided in developing System Dynamics.

The book also briefly traces our recent development on this planet in relation to one another and other living species, by considering how human behaviors have impacted the paradigms and ways people think. As many know, the wisdoms of *Tao Te Ching* can be interpreted in different ways, but there are striking parallels between passages that were written more than two thousand years ago, and important principles of Systems Thinking and System Dynamics that have been observed in the last few decades.

It is my hope that readers may find this exploration to be an enjoyable challenge which shows the positive impact that Systems Thinking could have on our global development as a species. The good news is that worldwide interconnectivity using computers and the Internet may help us to extend the powers of our minds to achieve new levels of understanding and rediscover what is important for all of us.

The Tao of Systems Thinking is divided into 9 chapters of passages and commentaries, followed by a translation of *Tao Te Ching* for those who wish to compare with the original text.

People may agree or disagree with ideas, but it is more important to think about and debate our roles as the inhabitants of this planet: for who we are and what we ultimately want to become.

Michael McCurley San Jose, Costa Rica, 2015

*Jay W. Forrester was the Director of the M.I.T. Whirlwind Computer Project during the 1950's, the inventor of magnetic core RAM memory, and the founder of System Dynamics.

*Lao Tzu is historically identified as the writer of *Tao Te Ching*, though knowledge of his origin is mixed with speculation and legend, as are other historical figures of antiquity.

This book is dedicated to the System Dynamics Group of UMG, Myanmar: Ei Ei Khaing, Khin Su Hlaing, Lai Zarni Khaing, May May Tin, Moe Thuzar Win, Nyein Nyein San, and U Soe Aung (2011-2012).

Thanks to Dominique, Donald Calvert, Harry Towne, and Phillip Reed for editing and reader feedback.

Thanks also to Michelle, Kitty, and Victoria who have worked on the Chinese translation in Beijing, China. Thanks as well goes to our translation and editing teams in Costa Rica, Hans and Frans Wittebol for the Dutch (Netherlands) translation, and to Sandra Bolivar and Margarita Guido for the Spanish translation.

This edition has been updated to match each passage to its related commentary for easier reading. Recent updates have been synchronized for both the print and digital versions of this book.

This is dedicated to the memory of Jay W. Forrester 1918-2016

Table of Contents

Chapter 1.

Our Universe

阴

Passages 1 - 9

1.

We all have a mother, and we know her name, but the mother of all things has no name. We call our mother: **Universe**. All things flow from this same source and are interrelated.

1.1—The connection between our world and the universe

Today we live in Gaia, a world which seems complex and difficult to understand at times, but we know where we have come from and what we are a part of. Having learned to manage small parts of our local world, we often compete with one another to obtain our wants and needs. Although our species has managed to dominate much of the planet we inhabit, Mother Gaia is relatively insignificant in comparison to our solar system, the galaxy, and the rest of the universe.

The boundaries of our understanding are established by the limits of our perceptions. Like flatlanders in a three dimensional world, our universe is a dynamic unity. What we usually perceive is linear, consecutive, finite, and disconnected, but all of these have the same source and are interconnected. [1] [2]

2.

What is manifested is generated by what is hidden. There is harmony in balancing opposites.

1.2—The harmony between apparent opposites

The macro-cosmos and physical manifestations we see in our universe are bound by unseen forces such as atomic forces, energy, entropy, and gravity. These create interrelationships between macro-cosmic and micro-cosmic systems and tie them together. Living systems are also interrelated in similar ways, but through different means, in unicellular and multicellular organisms, and their interrelationships are regulated as well through organic chemistry, renewable energy exchanges, and entropy.

The overall result of these interacting systems is a balanced harmony between forces which may seem to be opposites:

the centripetal force of gravity which binds planets to solar systems,

that same force which clusters stars into galaxies,

or the binding forces of much smaller sub-atomic particles in atoms and molecules,

the organic components which form cells and living organisms,

and the regenerative nature of life, which re-initiates as individual life cycles end. [3]

3.

A balance among all things is equilibrium. This is difficult to attain and also to maintain. Living systems, however, harmonize and achieve a natural equilibrium.

1.3—Balancing interactions

The resulting balances of constant system interactions may at times achieve equilibrium. This is difficult to maintain, since relationships between multiple systems depend upon shifting potential and available energy—and entropy (energy lost). So a constant factor in the universe is change.

Eventually, the balance or equilibrium of systems will be lost as entropy increases beyond the limits of available energy. These are principles (also known as the first and second laws of thermodynamics in physics) which apply whether we are talking about the life cycle of a star, man's use of limited hydrocarbon resources for energy and transportation, or the life cycles of multicellular organisms. [4]

4.

Once we use what we have, it cannot be used again. Levels of energy are common to all things, and flow from higher to lower levels. Organic systems, however, are constantly renovated.

1.4—The underlying constant of entropy

Entropy is the energy limiting constant of all systems. Energy always flows from higher levels of order to lower ones—from usable to unusable. Like heat, energy dissipates. The solar energy our planet receives from the sun dissipates back into outer space. The equilibrium between the solar energy received and the heat that's dissipated produces average global temperatures and climates. This, in turn, along with the existence of liquid water and an atmosphere of oxygen, nitrogen, and carbon dioxide, also supports the existence of hydrocarbon based life forms. Curiously, life recycles and regenerates what would otherwise be dissipated, by using solar energy.

All systems are regulated by entropy, which balances available supplies of energy and material resources, that are used and recycled, or become unavailable wastes. These systems function as long as more energy and resources are available. Some physical systems may change or break down if there isn't enough energy to sustain them, or as system infrastructures deteriorate. [5] Organic resources, however, are recovered and regenerated continuously, while the sun provides renewable energy to our planet. So natural energy systems in living organisms are constantly renovated. These natural processes have sustained life for millions of years.

5.

Space seems mostly empty and limitless. The more it extends, the more we find. There is an inner as well as an outer aspect to space.

1.5—Exploring inner and outer space

Much of the universe is empty space, but in comparative terms (to us anyway) it seems limitless, with a recent estimated radius of close to 14 billion light years. The more we study it, the greater we discover it to be. So far, we have been able to orbit our planet (fairly successfully) and make an occasional excursion to the moon, but those are tiny fractional distances in comparison to a single light year—the distance which light can travel in a year. We have yet to cross *one* of those light years. Comparatively speaking, this would be less than one centimeter of one hundred thousand kilometers. [6]

There is an inner nature to our explorations, which delves into the micro-cosmos as well. And as our investigations proceed, we uncover other layers of physical dimensions which regulate the subatomic particles of atoms and molecules, or organic chemical and genetic factors that regulate living organisms, which include us. Then for the realm of understanding and contemplation in our thinking processes, there is also an inner space within us as individuals.

6.

The essence of the source never dies. It extends like a spider-web into an interconnected network of all systems.

1.6—Consciousness and contemplation

Considering such tremendous distances, it may be surprising to know that the essence of everything in our universe is interrelated and unified. All of this is linked by the same web of interrelationships, of dynamic and cyclic interactions between systems. And although there's much to be discovered, there's much we already understand. [7]

We have become the first species on our planet to become sentient and aware of who we are. This has given us abilities to dominate and change our environment, though not always with the results we desire, or without unforeseen negative consequences. While we have learned to manage a small part of what we do in order to prosper and survive, we've not always done that effectively or well. There are other levels of awareness we have yet to master.

7.

What lasts or lives forever doesn't live for itself. Only by looking beyond yourself can you see everything else.

1.7—Becoming aware of a greater continuum

The essence of everything we perceive is the universe itself, which relative to us as living organisms, never dies. Life regenerates as part of the same continuum. A problem exists in the rather short perspectives we have of day to day existence in our own lifetimes. We tend to perceive things around us in limited time frames as individuals, who don't take much larger dynamic time spans into account, which we haven't yet fully understand. [8]

As we become consious of events beyond ourselves and our own limited spheres of influence, we also become aware of principles which extend through space and time. Thus it becomes possible to extend our understanding beyond the limits of individual perceptions through what we learn from others. Important knowledge never dies. It is passed on from generation to generation as part of a general continuum of human consciousness.

8.

Like water, flows between all things connect systems together. We make better choices when we understand how things flow and are interrelated.

1.8—All systems are interconnected

Flows within and between systems connect all things effortlessly. Flows interchange matter and energy through space and time. All systems work in this way, though we come to understand them through their different manifestations. Dynamicists who study these flows can create models to understand how systems work, project or simulate possibilities, and identify issues or problems which affect those systems. [9]

People may not make choices which are optimal because they don't understand how the systems which affect them actually work. Since certain systems are counterintuitive, some people may make decisions that produce undesired results. Dynamicists who understand how system flows are interrelated may be able to find the root causes of problems and resolve them. This doesn't always work in a practical sense because issues which affect the majority of us won't have the same impact on those who exercise control. [10]

9.

To keep the most we can get, we would have to stop time. This is a universal law: *learn to stop once you have finished your job!*

1.9—Selective distribution

In spite of the immense physical universe and our awareness of it as a species, men have chosen to compete for control of physical resources, and most people's lives are spent working, or struggling, in order to survive in economic systems that don't allow them to share the tremendous potential wealth that might have been available to everyone.

Capitalism and other economic models maximize profit-keeping motives for only a few select individuals. Businesses and corporations subscribe to and practice principles which are intended to maximize extraction of human and physical resources to benefit a corporate elite. Despite apparent distribution disparities from the start, those who practice this are rarely concerned about the inverted long term consequences of exponential growth. [11]

People who ignore this hope that the resources of our planet will somehow be infinite. At least this is what the elite seem to be saying—when they take or limit resources for others—*Eventually everyone will have enough resources if you follow our leadership.* The problem is that people who are unaware of the limits to exponential growth don't know when to curb exploitation and development until it's too late to avoid serious consequences. Should we follow them? In any case, there would be enough for everyone if resources were distributed more equitably in the first place. [12]

Chapter 2.

Balance

阳

Passages 10 - 18

10.

To maintain a balance that unifies your energies, you must be highly flexible. By adapting to the cycles within systems, you will discover their hidden virtues.

2.10—Understanding and leadership

Leaders who have real understanding of systems recognize that social systems can function in harmony because they're interrelated, like natural systems, with one another. Those who have this wider perception are more flexible, and learn to work without self-interest, because they care about others and realize that this is the best way to ensure our well-being as individuals and as a species. The best leaders have an essential understanding of how things work.

Those who understand the balancing relations of cyclic systems can govern with intelligence. They are not trapped by their own subjective experiences, but instead comprehend a wider scope and larger perspective in terms of space and time. And as a result, their decisions go beyond day to day concerns, seasonal considerations, or election cycles. They see humanity as a whole throughout time. [1]

11.

While material things have their advantages, what truly becomes useful is to discover their intangible relationships. We may often upset the balance of systems because we don't recognize that certain small variables can have great impacts in the long run.

2.11—Tangible versus intangible values

While most people focus their attention on the concrete materialistic world, a few understand deeper intangible relationships. This isn't surprising because people who manage superficial economic systems tend to separate us from the basic resources we depend on. This is more from intentional design than coincidence, because it causes more concern about symbols of artificial currencies, which fluctuate and are only worth what people think they're worth. [2]

While we build physical and organizational structures, what lies within them has greater value. This is similar to a parallel we might draw about containers and what they contain. What is within has greater value, but we must also consider what holds them. So value comes from inside and out—more from potential or purpose, than from physical reality.

12.

Pursuit of sensorial experiences can confuse the mind. One who is wise will learn to look within. Although we may depend on what is immediately visible and apparent, looks can be deceiving. What we see first may be symptoms of causes which aren't so easy to find.

2.12—What limits having more

Overwhelming the senses won't increase knowledge or experience, but might indeed mute or limit them. The same is true of growth or progress. Within what causes growth are the seeds of other feedbacks which will limit or stop it. Chasing after externals doesn't increase what people value, though some assume that wealth creates value and makes people valuable. Ultimately, wealth makes a few people more powerful and gives them greater abilities to control others, but not usually for reasons that benefit the majority.

Can we really possess the earth any more than our individual lives allow? To extend beyond what a few selfish individuals may control in their lifetimes, a small minority of individuals have created corporations which embody and consolidate corporate power beyond single lifetimes—thus perpetuating selfishness and greed on organizational and institutional levels, transcending individual life cycles. Although the value of this is questionable, governments nearly everywhere in the world have ratified the possession of private property in ways that deny direct access to resources for large proportions of our populations. Contrary to what most of us need or desire, the rights of elites to control resources, and possess or own more property, implicitly exclude the rights of others to do the same. [3]

13.

Failure can be an agreeable surprise. The humility of getting up again helps you to grow. Far from being something shameful or embarrassing, we should value every mistake we make or discover, as an opportunity for fixing an issue or problem.

2.13—Why systems fail and what makes them work

Failures with the systems we have created might produce important advances if we could change and learn from them. We would probably not advance or improve if we didn't fail at times or notice when we need to do things better. We should also recognize stagnant social systems which have frozen development and denied benefits to billions of people all over the world. These social failures have been institutionalized, but not without reason. Those who benefit from all this don't wish to lose control by sharing more equitably with others. [4]

We may be surprised when we make new discoveries about improving how things work. Our discoveries about systems go beyond our senses, beyond what can easily quantified or measured. By understanding these intangible forms, we can discover principles which govern the universe. Our species is subject to those same universal laws which require greater wisdom. Governing leadership requires empathy that many leaders no longer have. To administer our world, we need to care about our species as a whole, not as specific local or regional groups. This includes all political affiliations and religions.

14.

There are frequencies of forces we do not sense and yet they are all around us. Inaudible, incorporate, and without visible form, they continually flow through all systems. To know their origins is the beginning of wisdom.

2.14—Forms and structures

The forms and structures of many dynamic systems extend beyond what most people understand. A mathematical basis for them, for example, is calculus. These exist without physical forms, but they can be expressed mathematically. We may not see everything we try to discover, but physical evidence for systems is all around us. [5]

Knowing the Origin is the Tao of Systems Thinking. Many people try to dominate systems without a real understanding of how they work. Such attempts may fail because the behaviors of many systems are counterintuitive. Selfishness is not the beginning, but the end of wisdom. So those who are selfish may end up doing the opposite of what they think needs to be done.

15.

The wisdom of the ages is subtle, flexible, profound, and global. Those who study this know they can never know enough, and because of this they learn to understand.

2.15—The nature of understanding systems

People who are skilled at Systems Thinking have a wisdom which is subtle, flexible, profound, and global. [6]

—Such an understanding is subtle because it isn't selfish or subjective.

—It is flexible because it includes considerations for different options or outcomes.

—It is profound because it deals with root causes of issues, rather than their symptoms.

—And it is global because only global solutions will actually work.

16.

The supreme cycle is emptiness and return. Clear your mind first to find your destiny. Realizing that destiny is to find eternal purpose, and from that you can understand everything else.

2.16—What systems thinkers understand

Those who work with systems know that their understanding is never complete and can always be improved. Because those who are aware haven't learned everything, they can return to learn new principles which lie beyond their imaginations. And once these are discovered, new understandings will change how we work with systems.

To flourish, we must return to our roots and find relations which extend beyond time and temporal manifestations. To do social justice, we must understand systemics more completely. And only through justice can people govern effectively. [7]

17.

The best work you can do is to convince others to carry your cause and make it their own.

2.17—What governs us are not our own rules

Those who govern the least are able to govern best. People who govern repressively will be noticed and despised. Those who rule through fear are also defied. Such conditions cannot last. [8]

One who governs best shows or teaches others to make his cause their own. Thus when these people act it will be as if it were their own cause. They will say, "We have done this." And then they will act as if they freely chose what they are doing, without seeing the invisible hand that guides them.

18.

Those who abandon a general understanding of systems allow confusion to reign. Although they may seem intelligent and smart, the most they can attain is a paternalistic system without harmony.

2.18—Subjective intelligence

Many people depend upon their own subjective intelligence, which has no relation to Systems Thinking. More often than not, such thinking is relegated to selfish objectives and motives—based on individual survival modes—what social Darwinists call "survival of the fittest." In many cases, such intelligence is derived from desires for personal power or profit. Although people may seem "smart" when they show others economic, material, or social strategies, this limits the scope of wisdom for our species as a whole. Why?–because it tends to take resources away from the majority to benefit a minority.

Such people dominate and take advantage of others without considering anyone except themselves. Using mantras like "survival of the fittest" and "God helps those who help themselves," they've created elaborate control systems, often clothed with respectability as businesses and corporations, which extract benefits from workers, the same as any opportunist. This justification takes resources from others with complicated reasons or "rights" which most people fail to resist or understand. Thus paternalistic systems rule the day, and take advantage of economic disparities (called profits or earnings) which increase over time. This is called the creation of wealth. And while some people use wealth to become richer, what's left unspoken is that what creates wealth also creates poverty. [9]

Chapter 3.

The Great Design

陰

Passages 19 - 27

19.

A true way we can find a Superior Principle is to renounce learned behavior, benevolence, and justice. See what's simple and embrace the Primordial!

3.19—Education and conditioning

Learning to think goes beyond what most of us have been trained or conditioned for. Most educational systems prepare us to work in social systems we don't control. Before we can discover underlying truths for ourselves, we must shed the conditioned paradigms and mindsets we have been molded to follow. This requires us to question things first, rather than to follow without questioning. So we must renounce even what is apparently good or benevolent before we determine what to value for ourselves. [1] [2]

Simplicity is the root of more complex systems. It is easier to work with complicated issues by starting out with basic components and principles. From these we see how more elaborate structures emanate. Once we grasp fundamental principles, we will understand better how more complex things work and be able to troubleshoot issues when things don't function as we expect. This often requires us to forget what we have learned, and rethink things for ourselves—to go beyond *how* and understand *why*.

20.

What distinctions can we make between "good" or "bad"? Since when do we distinguish what we do because we fear what others think? Learn to think for yourself!

3.20—Learning how to think

People often make simplistic judgments about what is "good" or "bad" without understanding because of social conditioning. They react or respond from fear of what others will think, without thinking for themselves. This wastes cognitive and other human abilities:

—to have the ability to think without thinking,

—to have reason without the use of reasoning,

—to be passive when we have the ability to act,

—to add value to work which is taken away by others,

—to live under repression without resisting.

All of this happens when we consent and accept how we live without questioning this ourselves. What if we're the ones who are responsible for our own mental enslavement?

Do you think actively or passively? Learn to think for yourself! Be responsible for your own existence and learn to think critically. Not just to complain—you must learn how to transform your environment—first by focusing mentally within—and then the external, physical world will begin changing as well. [3]

21.

The universe seems confusing and imprecise. But within it are forms and structures which determine all that is real. You must also know this nature is within you.

3.21—Simplicity is the basis for complexity

The external manifestations of this world may seem confusing and imprecise because we observe phenomena, which contain mixed energy and entropy, of multiple intersecting systems with different dynamic relations—*all at once*. No wonder this seems confusing. And yet on a larger scale, we can see the natural harmony of the universe which is also reflected through living systems on our planet.

Why then, the apparent confusion? External manifestations come from simpler internal forms and structures. This is apparent even within organic systems of a single living human being. And even more amazing is what we know about the genesis of the human embryo—how it is conceived, develops, and grows. So from simple beginnings, we can see extensive development, also in ourselves, of much more complex systems. Thus we know that the nature which generates complexity is within us, but it has roots which come from simpler internal structures. [4]

Complexity is inherently unstable. Complex systems tend to break down more quickly than simple ones. This explains why so many of the complex systems we create have collapsed, and that most human endeavors do not last.

22.

Empty yourself, so you will be filled. If you have little, you will gain much; but if you have much, you will only become confused. Unity is a model that can be a source of completion for you.

3.22—Learning and thinking machines

Learning begins with a process of emptying yourself. Depending on what you know doesn't prepare you to learn anything new. If we are aware we know little, we are prepared to learn more. But if we have learned much already, we may be confused by new discoveries. Sometimes, we must forget and replace what we have learned—because what's new is contradictory and what we know becomes obsolete. This is also a defect of technological development, which makes learning become obsolete faster than when technologies are used for longer time spans. Natural developments tend to be more stable and enduring.

Dependence on technological development does not present us with lasting knowledge or solutions. In fact, today's "solutions" often become tomorrow's problems. Unity in nature and the universe is a better model that can be an important source of personal satisfaction and completion. What we learn from nature won't be surpassed or become obsolete, so what we know about our environment can lead to important understandings for our own development. [5]

Think of the first computers—these were technological marvels to be sure, but they filled entire buildings, cost millions of dollars, and were only affordable for governments and military use. Then mainframe and mini computers were developed for businesses and corporations, after which followed personal computers—the early desktop computers with CRT (television type) displays. All of these computers required costly materials, and used more energy, besides producing more waste. Recently, smaller laptops, notebooks, and tablets have been developed, often with the same or greater capabilities, smaller in size, but more energy efficient and less expensive. As this process continues, people learn to do more with less if they imitate nature.

The mental capabilities of a single human being are still vastly superior in many ways to even the best computers or thinking machines out there. [6] Are we effectively using the capabilities of our own minds? Learn to use your mind first—and then use the computer as a tool to extend your capabilities. Although computers can be programmed functionally to operate at great speeds, what do they actually *understand?* Machines are not yet conscious like humans, nor should humans try to imitate machines.

23.

Simple and tranquil words speak for themselves. Human efforts often precipitate violent effects without purpose. One who courts virtue, finds virtue. One who courts loss, finds loss.

3.23—Knowledge cannot be forced

The deepest knowledge speaks for itself. People often try to force things which naturally occur anyway. They do this to obtain what they think are additional benefits. This shows ignorance about natural abundant systems. Our world has resources which are potentially available for everyone. This has been true for thousands, tens of thousands, and hundreds of thousands of years because those resources are renewable. And yet we have strived, competed, and fought violently with one another to extract and dominate resources which were freely available, so a few could derive additional power by controlling them.

We shall see that those who strive to obtain more through violence court loss. And those who discover what is already available, gain both mentally and physically, because they understand that using force has only been employed out of selfishness or ignorance. Even in a world of diminishing resources, there would be enough for all who learn how to manage and share them. [7]

24.

We cannot tiptoe through life, but our own achievements won't last. There are other determinants we must recognize, which are stronger and will inevitably change whatever we do.

3.24—Determinants

Life would be empty without our own accomplishments, but these achievements, even for great endeavors, do not last. Other temporal forces soon erase what we've done. There are other forces of far greater scale and importance than the specific or individualized efforts of what we do—even when these seem relevant or important to us.

Greater determinants will change what we do, sometimes because we've been socially conditioned, or because other causes are fundamental to everything that exists, which will inevitably change us instead. The underlying nature of our environment is greater than the material resources we try to exploit or control. [8]

25.

There is a Great Design that is undefined but complete within itself, born before the heavens and earth. Its greatness extends far beyond our perceptions, but returns to be within them. Man is still bound by the laws of planet earth, but the Great Design has no boundaries.

3.25—Beyond perception

The immensity of the universe extends far beyond our mental and physical perceptions. And yet what is beyond us is still accessible, because what is undefined is part of the same continuum, whether we recognize it or not. Earth is part of that design, as we belong to it. So that immensity of space extends for light years that we can see in the stars, and yet it's still within us.

How could scientists like Albert Einstein probe the unfathomable depths of atomic physics to unleash the power of nuclear energy? And how could the instruments of astronomers extend our vision across billions of light years of interstellar space? While we're bound by the laws of our planet, our minds aren't physically limited in the same ways. [9] Thus we can sense the Tao of the Great Design in all systems, even beyond our own mental and physical perceptions. We are innately and intuitively conscious of that design in the Universe.

26.

One who travels should be well equipped. Those who act lightly may be separated from their own roots. Becoming agitated may cause people to lose control.

3.26—Making sense of subconscious experience

The cognitive abilities of our brains can be overwhelmed by our senses, emotions, and useless information. So it's best to be selective about what we learn, and conserve what is important. Those who are caught up by superficial aspects will be constantly agitated. But wiser men, who know what is profound and well rooted, are tranquil and serene. [10]

Focus on what is well based, and you will be assured of a solid foundation. This does not depend upon anything you can memorize or carry. It is within you and related to your universal sub-consciousness (and the Internet). Since nothing important can be taken away, it will never be lost. [11]

27.

A critical path follows a precise trajectory, when everything works as planned, using an effective guidance system. For this reason, those who are better prepared become teachers of those who are less so. This essential process leads the way.

3.27—The guiding principle of simplicity

The essential process that leads the way is the guiding light of Primordial Simplicity, called the Tao. This inner principle shows us that moving in the right direction leaves minimal effects in its wake and cannot be easily criticized. It shows us that good calculations don't always need a calculator or computer, that what's most securely bound is held without tying, and that no one is left out of a truly successful social system.

We need to focus on doing what truly makes us happy. This is quite distinct from pursuing materialistic goals to obtain what we *think* would make us happy—but actually don't. So we need to discover how to move in the right direction, because it's no use to move in the wrong direction, no matter how intelligent it seems. [12] This implies that there *is* a better direction in which we can move, for ourselves as individuals, and as a species.

Chapter 4.

Models and Government

陽

Passages 28 - 36

28.

To have a vision for a world dynamics model is to *be* a model for the world. Be humble. To become a source of knowledge for the world you must return to the Primordial. Simplicity is a key factor. The tailor who works best makes the fewest cuts.

4.28—Modelling world dynamics

There are a few people who model the world like Jay W. Forrester, but not many of them. When they do this, others are quick to react and try to negate what has turned out to be well reasoned warnings about the impacts of our behaviors on the planet. This may happen because some people are against any limits on their actions or behaviors, even when these limit the future for what others can do. [1]

When people become models for the world, other like-minded people will follow and respond. Even those who question them will carry forward a debate for the rest of us to consider. At some point, we will become aware when a critical mass has been reached.

29.

One who hopes to conquer the world cannot succeed. Our world is a sacred place and cannot be manipulated or dominated. For that reason—there is a time to move forward, and a time to move back—a time to increase, and a time to decrease. One who is wise avoids extremes and excesses.

4.29—The myth of global domination

Among those who think about the world are those who want to dominate it. Who hasn't thought about conquering the world at one time or another? People who try to conquer the world have certainly thought about it. That's why we are where we are—wherever that may be.

Will people who try to dominate the world actually prosper? This cannot be possible in the long run. For inhabitants of this planet, it is impractical to dominate extensive parts of it for more than limited periods of time—which are determined by rather short human life spans. There may be moments in which dictators may succeed—followed by others in which they fail. [2] As we know, there's a time to grow, a time to decrease, a time to live, and a time to die. All of these are governed more by system laws than by human beings.

30.

One should not attempt to govern by force because that force tends to revert against those who use it. What one wants is to protect boundaries, not to promote expansion. Excessive expansion precipitates collapse. What runs contrary to this principle will cease to exist.

4.30—Superpowers and the arms race

At one time in our not so distant history, two superpowers seriously competed to produce weapons of mass destruction which could annihilate nearly all living creatures on our planet—especially *our* species. Both super powers engaged in a global nuclear arms race, which proceeded rapidly until one generation succeeded the next. Then one of the superpowers collapsed, while even the other came to question what in the world it was doing. This was not before those powers—and other nations who joined this collective hysteria—had produced more than a hundred thousand nuclear warheads—more than enough to destroy all cities of any importance inhabited by human beings on this planet. [3]

At some point, a few more level headed leaders were somehow able to step away from this massive insanity— but only after two cities were destroyed in seconds at the end of the Second World War. It was not until an entire generation had passed that people realized they had no inclination for destroying more cities with atomic weapons, even as sophisticated weapons development made it much easier to do this. [4]

Since then, a few leaders have made lukewarm gestures for disarmament, while others still hold onto those relics of mass destruction, perhaps out of fear or nostalgia, or just because they can—which isn't far from state sanctioned terrorism—though we are advised that for our own defense, we have the capacity to destroy ourselves.

Such is the logic of armaments and war.

31.

Sophisticated weapons of war presage calamity. To rejoice in military victory is to rejoice in the killing of human beings! This cannot translate into prosperity for our species. Sooner or later, every war is a massive funeral.

4.31—Global warfare

During the 20th century, like other centuries before it, different nations of our species fought in several wars, including World War II, which cost more than 50 million lives. After more than 100 million casualties in two World Wars, we came to realize that the people who became our enemies could have been our friends. This wasn't much different from what people learned during the 19th century, but had somehow forgotten. Friends became our enemies—then these enemies became our friends—and so on. Every generation is much like the next, since people end up fighting each other and then make peace later to become friends. [5]

Those we fought against were not very different from ourselves. In fact, genetically, they were nearly identical, but outwardly they seemed different. So people dehumanized others in order to kill them. This came from the error of believing there were superior groups of our species, who somehow had the "rights" to control, conquer, or wage war with others. And this came from the leaders of their governments or people who controlled them, not from the general masses who inhabit the planet. [6]

For the millions who died as a result of fighting among ourselves as a species, these wars were massive funerals. For every country that celebrates its own memorial day, there are more important lessons to be learned than valor and patriotism. Our world does not exist to be dominated or controlled, and any efforts to do this are soon extinguished. Instead, our world is meant to be shared and appreciated. There's no need for warfare to do this.

32.

Out of simplicity comes complexity. The Universal lacks a name, although what is Primordial, self-organizing, simple, and small, it is in no way inferior to everything else, and all things are subservient to it.

4.32—Organizing principles

The organizing principles of our universe have simple fundamental roots. All that exists is ultimately energy, matter, space, and time. While these principles are simple, what comes from them in endless combinations is more complex. Just as an entire language comes from a few dozen letters with vowel and consonant sounds, we will also find that most of the physical universe is based on a few dozen principles. [7]

To understand how things are organized, we should look for root causes. Complex systems are based on simpler components. We should simplify and look at basic relations before we attempt to work with more complex ones. Problem solving works in the same ways. We need to identify the root causes of problems before we can solve them. If we deal with symptoms instead of causes, it's unlikely we will find lasting solutions for the problems we face. It is also unlikely we will resolve issues unless we recognize when our own participation produced them in the first place. [8]

From such reasoning comes the saying: *tomorrow's problems come from today's solutions.* Although this sounds pessimistic, it's often true.

33.

One who can know and conquer oneself is truly powerful, who knows when abundance is enough. Anyone who follows the means of systems, is someone with constant purpose.

4.33—Expanding awareness

A higher value of knowledge is based on individual perceptions and our greater awareness as a species. Although we prize individualism to a great degree, our cooperation as a species is impaired when we are reduced to primitive levels of competition. Inter-individual competition is useful for those who control us, because we tend to compete more with each other on similar levels, and pay less attention to those who exercise control on higher levels through social, government, or corporate domination.

Those who practice self-control, and free themselves from the controlling influence of their own egos, will generally become more aware. Systems Thinking provides a framework for thinking socially and globally. As people become aware, they will see reasons for cooperating with others for more than money or material possessions. Accomplishing this is difficult, however, since it's easier for people to accept what they know, than to anticipate what they can accomplish in the future. Creative minds, however, have always been capable of doing this, and will be interested in exploiting these potentials. [9]

34.

The universal means is like a powerful current that feeds us but isn't consumed. All things return to it as to their homes.

4.34—The underlying balance of global energy and resources

The universal force is a powerful field or current of energy that surrounds us. Although we may feed on this current, it is self-renovating and never consumed. Global systems are cyclic and regenerative. This contrasts sharply with how most people think, whose perceptions are limited by linear, consecutive time frameworks which assume that material resources are consumed, will run out, or be lost. This shows we're limited by economic perspectives we've learned in order to survive.

While there are limitations to non-renewable resources, this paradigm is taught to us by those who dominate resources which might otherwise be accessible for everyone else. Their power comes more from their abilities to limit resources and control others, than from making those resources available, because this would destroy their power base. [10]

In spite of this, we sense an underlying abundance of our planet's resources. This quietly reinforces the principle of *enough* if we could learn how to share. In reality, we belong to the earth more than it belongs to us. The earth existed long before we appeared, and it will probably exist long after we have finished our life cycle as a species. This implies that ownership is more of an imaginary issue—what we believe or think as an agreed upon game—not a prerequisite for our survival. Money divides us from the resources we use. [11]

35.

Those who have learned about modeling can simulate many things. They can test them without harm and can model lasting effects, though these haven't yet been tested in reality.

4.35—Creating models

People who create and simulate models are able to check behaviors and possible outcomes of systems they can replicate from the real world. Modeling is simply another way of thinking. We all have mental models of one sort or another, although some of these are often not well thought out or clearly expressed. A system dynamics model may not be significantly better than other mental models, but it's easier to validate and test because it can simulate a wide variety of behaviors and is expressed using the language of mathematics. Now that personal computers are widely available, more people can create simulation models and observe their results. [12]

Models can be tested without risking real resources that may become scarce or unavailable. In addition, a model can be tested without serious consequences, while applying drastic real decisions could have impacts with unpredictable results. By testing multiple possibilities, modeling may help us to expand our horizons, not because what we discover will be perfect, but because we will have considered more possibilities and scenarios. [13]

36.

Here is a subtle wisdom of life: there are weaker forces which have leverage over stronger ones.

4.36—Simulating outcomes

People who calculate on the basis of strength, look for what is impressive. That is an error of judgment. What we see as overt power depends upon simple origins. We'll be unable to move forward without clearly understanding those origins and returning to them.

Those who are wise realize that weaker forces have leverage over stronger ones. This may not be immediately apparent, but consider the hidden power of atomic physics. What's least apparent is the most powerful of all. Who would have imagined that nuclear power within the atom is thousands of times more powerful than the chemical bonds of matter we have known? The power of chemical explosives has no comparison to that of nuclear ones, and yet we had no idea about nuclear energy until the 20th century. [14]

Leverage isn't what most would think. A real understanding of underlying principals allows us to do things we never would have conceived of—with equally unimaginable consequences—unless we can somehow be prepared. Knowing the technologies involved isn't enough, as we can see from our behaviors after we discovered and started to produce nuclear weapons. For that reason, people mindlessly produced weapons without realizing what they were doing (out of fear) until they gradually became aware that their ability to create nuclear explosions was quite distant from an actual desire to deploy them and destroy everyone.

Awareness goes beyond mere consciousness or knowledge. It implies deeper perceptions than most people normally achieve—though this is what we hope to attain as a species. [15] Surprisingly, there are small signs we're moving in the right direction. The most important of what we learn may resonate until it becomes a common cause for all of us.

Chapter 5.

Dynamics

阴阳

Passages 37 - 45

37.

In a self-governing system, all dynamic aspects develop by themselves. If a system tends to oscillate, there are time delays which establish the frequency or period of oscillation. When these delays are moderated or damped, the system stabilizes again.

5.37—Cyclic or oscillating systems

Global systems tend to develop by themselves. As systems interact and evolve over time, they tend to oscillate and establish periods for repetitive or cyclic changes of economic, social, and political behaviors, or cycles of growth and collapse. People might see these clearly if they were more aware of long wave oscillations.

Economic systems grow and decline. Empires expand and collapse, while most democracies shift or switch political control between opposing parties. People with limited perceptions are unaware of how long term changes and delays in impacts occur over time. Even corporations are subject to cyclic changes. Those who are aware will understand how global systems develop and change. [1]

What does this tell us about "sustainable development"? If sustainable development is simply reduced to growth, no system is capable of growing indefinitely. At some point, growth converges and stops, no matter how successful a system appears to be. If sustainable development refers to a self-regenerating system with a constant inflow of energy, then such a system is sustainable. But development based merely on growth will end. Our planet is an excellent example of sustained natural development, since living systems have developed and been sustained for hundreds of thousands or millions of years. Solar energy serves as the regenerative power source for all living systems on our planet. Even our own recent exponential technological development has been powered by fossil fuels, which are concentrated forms of solar energy. Those who scoff at solar energy compared to more readily available fossil fuels are ignoring their source. [2]

The desire of our species to create order and dominate the world, creates corresponding amounts of disorder, waste, contamination, and chaos. While we focus on short term benefits for what we can accomplish, people often pay no attention until consequences become too obvious to ignore. Thus we create cycles of order and chaos, war and peace, tumult and calm, love and hate, until we recognize the patterns from our actions in the systems we're a part of. Then we'll become more aware and moderate our behaviors. [3]

All living systems have life cycles. The artificial systems we create aren't very different from natural ones. We can learn more by studying natural systems.

38.

Maximum growth is not a virtue, although it is an orientation for many. This ritual for maximum growth isn't the answer, though some try to achieve it through force. The external is but appearance. What is within systems is more important.

5.38—Our addiction to growth

Although people have learned a few basic concepts for how systems work, our economic systems are based on an addiction to material growth. Most business models promote growth as a rule. And some people will do almost anything and even resort to violent means to grow wealthy. Moderate goals for sustainable development are often only disguises for unchecked growth. Natural systems include growth cycles, but when biological organisms reach a certain size, growth is automatically programmed to stop. Corporate business growth and desire for monetary gain, however, have no built-in braking system—no defined limits or boundaries. [4]

What results do our behaviors have for our global development as a species? The economic distortions we create by producing sporadic and uncontrolled growth cycles, tend to impoverish people and benefit only a small minority. Economic growth tends to reinforce this effect so that greater wealth is concentrated in the hands of fewer people, while large proportions of our populations lose access to basic resources. If we consider the total available land area of our planet, we can see there's enough potentially available land to distribute property more equitably for everyone who inhabits the earth, regardless of where they live. [5] We know, however, that property isn't evenly distributed and many people who have been disinherited of their rights to own property live in slums, shanty towns, and *favelas*, with no share of real property—billions, in fact, have no landed property at all.

This may not have been the intended purpose of those who have become richer, but it's a logical result of their actions. And those who gain aren't complaining. The links between wealth and poverty are inversely proportional.

Another facet of uncontrolled growth is represented by the rise and fall of empires. Historical cycles show us how global power centers have grown, developed, and then collapsed, limited to an important extent by the life spans of human beings, or transitions from one generation to the next. [6]

People who deal with systems or try to manipulate their effects without clearly understanding underlying causes, only manage parts of a greater whole. They cannot understand internal mechanisms unless they have a more complete understanding of systems. Today, the underlying basis of Systems Thinking is System Dynamics. Mathematics is a fundamental language for System Dynamics, and computers are used as calculators to "do the math," while people experiment with different scenarios to interpret how systems work. Users of different types of software are gradually integrating common definitions, functions, and terms. [7]

Generally, the mathematical relations in System Dynamics models express more clearly what we're doing to ourselves. Although those models may not show what most of us would want, they could also be used to demonstrate new possibilities.

39.

All things are what they are by virtue of an interconnecting unity. Humble origins for systems are the roots of greatness.

5.39—*Common causes*

All fundamental relations in the universe come from interactions between matter and energy, through space and time. These are the basic elements and relations of all systems. The origin of what exists is relatively simple, and from it comes the complexity of everything else. [8]

Because of this, we can understand complex systems as long as we recognize that everything is interconnected. This same principal applies to our existence as a species. Although there may be different human behaviors, there are no important differences between people as a species. We are fundamentally the same, regardless of where we live. That is, we're almost genetically identical, as well as linguistically, socially, and physically similar to one another. [9]

Our economic, political, religious systems, and languages are more similar than different. We behave in the same ways in times of crisis. We use the same technological tools, and fight, love, raise families or wage wars in ways that only humans do. It's surprising that people are not more aware of this—but perhaps they *are* aware and choose not to be conscious of it.

40.

The cyclic movements of the universe consist of means for their return. All corporal things in the heavens were born of incorporeal ones.

5.40—Matter and energy

The entire universe is made of interrelated cyclic systems. Origins of energy and movement return to their source. In a simple sense, our world is made up of interacting opposites. As Einstein's equation $E=mc^2$ shows us, all physical things in the universe are equivalent to energy. So matter and energy are counterparts of one another. All that physically exists comes from incorporate energy. Space and time are also counterparts of the same continuum. [10]

Einstein's formula can be used to calculate equivalence between mass or energy, and scientists are gradually learning to make better use of this principle. In the past, however, what we learned was applied for military or destructive uses. So we learned how to make nuclear weapons before we moved on to harness nuclear energy effectively for more peaceful purposes. [11]

Many of our wars have been struggles to control material or energy resources.

Other important technological developments, however, have come from military applications as well. History demonstrates that in spite of our mistaken fits and starts, we're gradually learning what we need to attain awareness as a species through understanding the unity of opposites. Or at least we should be. [12]

41.

The study of systems must be practiced with diligence. Otherwise, the practitioner will oscillate between faith and incredulity. People who study without values will laugh at the "absurdities" for everything that seems apparently to be something else.

5.41—Understanding systems

Those who study systems should learn and practice carefully. Otherwise, they may become confused or doubt what they learn if they don't fully understand the underlying principles of what they're studying. Why is this so? Many systems are counterintuitive and defy the natural intuition people use to understand them. People with superficial understandings may be caught by apparent contradictions, and come to believe it's absurd to follow training in Systems Thinking. [13]

Such people may pay attention to those who have attacked System Dynamics and come to accept surface arguments against it. This might not be a result of their own considerations, but instead come from accepting the ideas of others, causing them to question their own thinking. This is a sophisticated form of unconscious social conditioning, when people accept the arguments of others instead of thinking critically and learning to examine evidence for themselves. And yet it's highly possible in a world where people are employed by time-sensitive corporations. People don't have time for themselves, especially not even to think. [14]

Systems Thinking isn't an unusual or a specific target of corporate culture. Nearly all forms of independent thinking would naturally be discouraged, except for those that fortify materialistic economic systems, which people are part of, to compete, survive, (or thrive). Points of view which aren't economically viable are thus "a waste of time." The next question might be: what is time for? For those who follow this fallacy: *time is "obviously" money.*

42.

The balance between opposites, between Yin and Yang, is a vital harmony that has spontaneously generated innumerable systems. By understanding their relationships we will also discover how we can gain from what is lost, or lose from what is won. And from this an important lesson is learned: *any attempt to provoke violent change or disrupt a system will also end violently.*

5.42—The paradox of human competition

Opposites and paradoxes for systems thinkers are not contradictory, but rather are balancing forces within systems that lead to harmony. This is the opposite of what most people naturally think. If we examine the history and philosophy of thinking, the principles of opposites for the Tao and dialectics as a synergy of unifying forces between opposites, have been common knowledge (to some thinkers) for hundreds or thousands of years. Accepting this, however, hasn't been easy for people who haven't been taught to think and tend to subscribe to single sided issues, without considering wider viewpoints which require using more energy to think than the minimum. [15]

Dialectics was also a key paradigm of Hegel's philosophy or Marxist thought, but has not always been applied in ways that were practicable or true. This is because socialists and other thinkers also mixed wishful thinking with their politics in order to sway the masses. This leads people to expect more than can actually be done, which also led to the downfall of political systems, when leaders who preached one thing, then practiced something quite different than what people had hoped for. [16]

Many democratic systems have done the same or worse, by pointing out that individuals should actually be free, but then legislate so excessively that their populations are over-burdened by sustaining corrupt and inefficient economic and government systems.

Many governments still use this paradox by creating expectations for their populations, but then handle things quite differently because there are time lags between promises and delivery. Somehow, governments don't represent the interests of the majorities for their populations, but they do manage to serve the interests of smaller minorities who control resources—*and governments*—regardless of existing political systems. [17]

Competition between different governing systems has produced wars between human beings that are questionable for many reasons, except for a few who expect to benefit. These few resort to violent means to obtain their objectives by mobilizing the masses of their populations in armies to wage wars. They don't care if in the long run, everyone loses more from wars than could ever be gained, as individuals, as citizens and patriots of the countries we inhabit, and most of all—who we are as a species.

From this we have learned and known for hundreds of years, that we lose most from what we seek to obtain through violence, that those who lose may indirectly win, and those who achieve victories through warfare—often fall later to violent ends. [18]

Even on a smaller scale, what sense is there to see young men cut down in the prime of their lives in drug wars between conflicting narco-gangs? What would the people who were killed in these conflicts choose if they had a second chance? There's no glory in war—only funerals—and nothing is won honorably if it's taken from somebody else.

43.

Flexible systems adapt more easily than rigid ones. There are advantages to stable systems where nothing else has to be done. Such systems are self-sustaining.

5.43—Flexible systems

Flexible systems change more easily than rigid ones. Not all democratic or socialist systems have succumbed to collapse. A few political systems have survived or thrived beyond all expectations—not because they were good or bad, but because they became more flexible.

China is a prime example, but we could also consider Germany, Japan, and a host of other countries, including the United States after its own civil war. These countries, even after being seriously damaged or nearly destroyed, have risen again from ruins and ashes like the Phoenix to become stronger than ever. Empires that have collapsed, at some point, became too rigid and were no longer able to adapt. The fact these empires could exist at all suggests that they were at first dynamic systems, which expanded and encompassed other systems, and were once more flexible than they became later on.

Rather than taking powerful nations as examples, we might focus, as well, on smaller systems which are quietly successful in their own ways. Canada, Costa Rica, Denmark, Norway, Sweden, and the Netherlands come to mind as countries which have tried to offer a higher quality of living for their populations and have succeeded in different ways. [19]

It would only be fair to point out that all countries could use their resources to improve general living standards if government leaders decided to do so. Ultimately, this does not depend upon which country people live in. It depends on the will of people who live in countries everywhere to make common decisions. Wars cause unnecessary wastes through destructive forces. What would happen if the resources used for wars were employed instead to create a renewable energy base, improve living conditions, and offer better education, employment, health care, housing, and other social benefits for everyone? It is sad that for most of us, however, this is only a dream.

44.

Placing too much importance on anything will distort its value and increase its overall cost in the end. Knowing when to stop growth can avert later dangers that will eventually alter the life cycles of systems—including our own lives.

5.44—Obsessions

We have a tendency as human beings to become obsessed with what we focus our attention on. This tends to distort values and increases the overall costs of what we value, with negative effects and consequences.

Unfortunately, we lose sight of an overall balance, when we focus on specific targets that interest us, while ignoring negative side effects. The possibility of *more* interferes with our basic knowledge of what is *enough*. This is not surprising because people have been conditioned to think *more* is *better*, even when it isn't.

If we recall that previous escalation of weapons during the Cold War, the United States and the Soviet Union fell into this trap when they entered into an escalating Arms Race. This caused both countries (along with others) to devote immense quantities of resources that were dedicated to producing more than 100,000 nuclear weapons. Although the Cold War employed millions of people, both countries could have invested their resources more wisely with better direct results for their populations. [20]

At the same time, both countries entered a space race which put men into orbit around our planet, placed the first men on the moon, and produced satellites, which now handle global communications. Defense industry computers and communications also evolved into personal computers, cell phones, wireless technology, and the Internet, which people now use regularly nearly everywhere. [21]

It is interesting that over time, people gradually come to recognize what is important, while previous harmful obsessions become a thing of the past. Although more highly destructive atomic weapons were developed, such as hydrogen bombs (except when dropped in an accident near Palomares, Spain), they have never been purposely used since Nagasaki against real cities as targets. We've discovered that nuclear accidents can have toxic environmental impacts for hundreds of years. Round-the-clock flights with nuclear weapons (and corresponding accidents) have ended, and submarine voyages (with occasional incidents and accidents) carrying nuclear weapons have decreased. [22]

Although we're not immune to wars in this century, we're becoming aware that wars aren't generally beneficial for the majority of human beings. And we now have more complete historical and documentary evidence of devastation caused by previous wars. We may not be able to entirely control our nature for destruction, but we're certainly more aware of it.

45.

Even what is perfect may not seem so, but its application is inexhaustible and not expended. As movement overcomes the cold, a state of rest dissipates heat. The norm is equilibrium.

5.45—Naturally sustained systems

Consider how the systems around us are perpetuated without apparent intervention. This may seem somewhat strange since most people think that human work or activity is necessary to sustain everything we have created. Living organisms in natural ecosystems, live independently from human economies and technologies, and have no need for our intervention or assistance.

Life on this planet has existed for millions of years without us. We aren't an essential component to life on planet earth, even if we have created complicated infrastructures which require our constant attention. Most living systems are self-sustaining and replenished by natural solar energy. These don't depend on any special activity from human beings. What they require is renewable energy from the sun, which we ultimately need from Nature and our environment as well. [23]

Unfortunately, we have created fossil fuel based societies which are dependent upon resources that will eventually be depleted. Our tendency to use energy fuels produced many of our corresponding technologies. First came wood based energy systems. Then came coal and technologies that were derived from its development. Finally, fuels from petroleum based or other fossil fuel derivatives, including natural gas, have produced newer technologies we observe today. These have forced shifts in our energy base as we move from depleted energy resources to other alternatives. [24]

Fossil fuels, as concentrated energy sources, have given us tremendous potential to expand our growth as a species over the last 200 years. This hegemony of fossil fuels and our domination as a species will be short lived, however, unless we soon become successful at tapping renewable energy from the sun. To do this, we need to create large scale regenerative systems that will provide solar based energy for everyone. Easier said than done. [25]

Otherwise, we have entropy as a negative motivating force that will require us to change. As the EROI (Energy Return On Investment) decreases over time, requiring more cost and energy to produce every kilowatt or horsepower of energy we use, it will become relatively more economical and necessary to explore new alternatives as it becomes more expensive to use the traditional fossil fuel technologies we already have. It takes energy to produce energy. At some point, we must replace those technologies with newer ones that will become more effective, renewable, and efficient. [26]

If we don't change voluntarily, we will be eventually required to do so anyway—not by human laws or government, but by the laws of Nature itself.

Chapter 6.

Understanding

太陰

46.

When we understand the world, we will live in harmony with our environment. But in a world without understanding, the earth's resources are consumed to build weapons of war. There's no greater calamity than not knowing the principle of *enough*, and no greater blessing than knowing that enough will enduringly last.

6.46—Tragedy of the commons

We compete with one another for resources in this world we inhabit. For that reason, we consider material resources to be more important than they really are. Ironically, because some people want more than enough—there will never be enough for the rest of us. In any case, those who want more than others act as though there's never enough by taking more than others, thus creating a self-fulfilling prophecy by making resources scarcer (and more costly) for everyone else. Although this sounds like good economic sense, economics is nothing more than a set of symbolic system conventions that human beings have created—not the real nature of our world's resources and ecosystems.

Motivation for this fallacy of human ego can be reduced to two basic maxims:

I want more.
There is never enough. (This produces infinite need or demand)

This line of thinking produces what we've come to know as the Tragedy of Commons. Although those like ecologist Garrett Hardin have used contemporary terms, we have been aware of this for centuries. The problem is that those who are selfish have discovered effective ways of controlling what they take, by creating an illusion that other people must depend upon *them* in order to succeed or survive, rather than the actual resources required. [1]

This has the dual advantage of making those who grab more resources rich, while giving them rights to control resources that would otherwise belong to others. Those who no longer have resources must work for those who took them. And it's useless to debate whether people should actually control what they already possess. There are laws to protect ownership. [2]

A mute-point, because we don't practice it, is that there would be *resources for everyone* if only we could understand and apply the principle of **Enough.** When people consume and use resources like there's no tomorrow, it may be because they fear what they might lose. And what would people lose or gain if we learned to share more equitably? We don't know the answer because we've rarely tried to be less selfish or created political systems that are economically equitable. Since we have divorced economics from politics, when we try to combine them, the results have been disastrous.

So it seems logical to let existing inequalities persist, and adjust to them until we are forced to do otherwise. [3] There's no greater calamity than not knowing when we have enough, and because of this, millions of people have been forced from their homes and been murdered or starved to death. This is the Father of all wars. There's no greater malice than coveting the resources of others, when that leaves people without their rights to survive.

Although we reassure ourselves that we're not responsible for the calamities that affect others, it would be naïve not to recognize that we've been willing or passive accomplices of those who have enslaved or murdered people—no different from ourselves. [4]

We will only have enough when we completely understand what sufficiency really means for our species. That awareness hasn't been reached yet. As we are rapidly consuming fossil energy resources without taking fuller advantage of opportunities we have to create new infrastructures for renewable resources, we demonstrate that as long as we remain egotistical as individuals, we will not become self-sufficient as a species.

47.

The wisdom of modeling systems allows us to know the affairs of this world without leaving our homes. We can know the behaviors of how things work without even glancing out our windows. A wise one knows, without having to act, through simulations.

6.47—Planning

Human beings, unlike other species, have learned to plan. We are able to take from imagination—what's in our minds—and convert this into reality. At the same time, some people are aware of negative impacts which will affect our future. Those who warn about the future are often not responded to favorably. And yet, we cannot anticipate future results with any reasonable accuracy unless we consider both the positive and negative feedbacks of our actions and behaviors. [5]

For future events, we have discovered a more exact language in mathematics. By careful study and observation, a few people, such as astronomers, can accurately predict cosmic events and model different types of dynamics. There are many types of modeling, some of which are more respected than others. Modeling exact events is highly valued because the measurements and mathematics for these are quite accurate. Modeling for System Dynamics, which is related to Systems Thinking, often contains many variables used to show possible outcomes, but the dynamics for events may change before they occur. So variable outcomes are not easy to predict accurately. Modeling different scenarios, however, can be used to show outcomes people might prefer or want to avoid. These provide feedbacks which allow us to take actions that may change future events. [6]

Why should we examine outcomes we don't want? If modeling shows alternatives to make changes or improvements, we can take advantage of opportunities or awareness to avoid possible pitfalls. It isn't necessary to field test or try every option or error. Instead, we can model possibilities and eliminate nonviable options by simulating what's more likely to work. Then we may discover why certain options work, while others fail. And doing this doesn't require that we leave our homes or field test every alternative if mental rather than physical activity could help us to identify workable solutions. [7]

48.

Most learning consists of accumulating knowledge, day by day, but using Systems Thinking reduces it to simpler forms, day by day. Ideally, the best model to follow is the simplest one, though that doesn't mean we'll be left with nothing to do.

6.48—The growth of knowledge

The problem with learning and all general knowledge is that it tends to multiply at an exponential rate. Our ability to process and retain information doesn't correlate to this. If we don't discriminate about what we learn, our brains will soon become overloaded with unessential information that blocks further input. We also forget, and have natural internal programming to forget. [8] This is an essential process as well.

While it's important to process new information as we learn, we must also be prepared to reduce or discard unneeded information that isn't required. People who are aware of this, such as those who have studied the Tao or Systems Thinking, do not simply acquire more knowledge. They select and keep what is important—and discard what is not.

More than 2,000 years ago, followers of the Tao were aware of this and learned to reduce, select or discard whatever knowledge that was unnecessary—before the existence of personal computers and the Internet. What they did probably makes more sense now, than it did before electronic computer technologies were introduced. While this may seem obvious today, it probably made less sense to people who were unaware that computers would be invented and we would be inundated with information. People now *have to be* selective about what they learn, and eliminate whatever is unnecessary. Otherwise, they would be unable to find what they normally need, even for day to day activities.

We need to find and keep what is essential. The rest is often unnecessary or unreliable.

49.

Wise people realize that their interests lie in the interests of others. They are open minded to new ideas like playful children. One who is wise is truly free to think about and recognize the truth.

6.49—Debating our use of a world model

During the 1970's, Jay W. Forrester made a world system dynamics model, and used it to warn of certain predicaments or issues mankind faced if people did not moderate their use of natural resources and share them more equitably.[9] William D. Nordhaus, a well-known Yale scholar, criticized Forrester's model and warnings, and claimed that these were inaccurate and incorrect. He further stated that there were no such real limits to resources. [10]

Time would test the works of both men. Forrester stated that if we do not moderate our use of this world's resources and find other alternatives, we will face serious consequences that could curb our growth and development as a species.

Nordhaus contended that Forrester's methods used computer calculations and data estimations which were inaccurate and misleading—and he made several arguments intended to disprove what Forrester had said. Computer science was still in its infancy, and there were no personal computers at the time for people to confirm or check what the two had stated (Forrester as a scientist, and Nordhaus as a scholar)—especially since they referred to future events. [11]

Had Forrester simply created a computer program to support what he warned us about, or was there validity to what he said?

In summary, the arguments of both men are as follows:

1) Jay W. Forrester created world computer simulation models to show the impact that growth and using resources would have upon our activities and the environment. He pointed out that there were limits to natural resource reserves and we would suffer negative impacts on our species as a whole, if we ignored or refused to take corrective actions, and instead, continued business as usual by exploiting resources at exponential rates of growth. [12]

2) William D. Nordhaus criticized the modeling methods used by Forrester and pointed out that it's possible to create models to support nearly any point of view, thus rendering them unscientific and invalid. He also claimed that Forrester was simply restating an older argument from Thomas Malthus (related to population growth and the food supply) which didn't take into account factors of human ingenuity and innovations that have helped us to solve similar problems in the past. [13]

The results after several decades are interesting. Both men were right to a certain extent, but the future was different in ways that were difficult for them to anticipate as well.

1)   It would be unlikely for us as a species to accept collapse without doing anything about it (Nordhaus). We have historically adapted before.

2)   Innovations in technology might be possible, but social and political will to implement major changes would be unlikely or difficult to achieve (Forrester).

Events that went beyond expectations:

—Mainframe computers were replaced by personal computers, so it became easier to check System Dynamics models than Nordhaus anticipated. Besides this, Forrester had provided multiple scenarios, and simulation models would evolve and be improved over the years that followed. Today people can check models with personal computers themselves, but 40 to 50 years ago, only a few mainframe computers were available. Using mathematics provides a language that people can replicate, work with, and verify.

—New technologies for processing petroleum and unexpected discoveries of natural gas and other resources went beyond the conventional energy scarcity scenarios anticipated by Forrester, which fueled new expectations for growth or recovery, with mixed results.

Predictions which were accurate:

—Concerns about absorption limits for anthropogenic products such as carbon dioxide, CFC's, toxic metals, and other contaminants have demonstrated negative measurable global impacts on arable land, agriculture, health, water aquifers, and species of other living organisms.

—Many of the results simulated in the World models from 40 years ago have been confirmed and observed, although not always in ways which were expected. [14]

—The possibility that people wouldn't take major steps needed to respond and implement major changes was also considered, and this is where we find ourselves today. Inaction has been the easiest alternative, although there have been limited attempts to reduce the impacts of our actions on the global environment.

Conclusion: There have been interesting changes in technology and human awareness which have set the world stage for future developments. So to paraphrase what one prominent American has stated: *Never waste a good crisis as an opportunity for change.*

We have begun to witness a series of crises which may give us the opportunities we need, but will we improve by taking advantage of them? [15] Paraphrased quote by Rahm Emanuel in "In Crisis, Opportunity for Obama"

50.

Life and death have multiple companions. Those who contemplate or fear death will have a lot of company, but that's because they are struggling for life. And yet there are some who pass through great difficulties almost effortlessly untouched. That is because they allow for nothing less than life.

6.50—Live or let die

Although we may not notice, we make choices between life and death every day. We can choose how we respond to every situation and in most of these there are choices we can make for life… or death. Creative choices are companions of life, destructive choices are companions of death, and entropy is the end of everything in frozen stasis or heat death.

Our response is important because we're well aware of what the end will be for us as individuals. So in simple terms:

—*How we live is what matters if death is a foregone conclusion.*

Why are some so vibrant and full of life, while others struggle with living? Those who live fully accept nothing less than life, and those who experience the opposite are obsessed in some way with death.

Living requires the fusion of creative energy.

Death releases the fission of destructive energy.

But creation can also be a synergy of creative destruction in which what's new replaces previously existing counterparts. These aren't really positive or negative, though people may interpret them in simplistic terms. [16]

It is an idealistic fantasy to say that life is the ultimate goal for everyone. Some find a dark gothic comfort in death as the ultimate form of repose or revenge, since no one can persecute anyone from beyond the grave (but then religion steps in to rectify that with an afterlife). Suicide bombers and other terrorists have lost or obliterated any joy left in themselves for living, to the point that they're willing to sacrifice life by making a "statement" in committing suicide or killing other human beings.

Then there are also natural disasters which occur unexpectedly without any human intervention, and of course other disasters which are combinations of both. These challenge our understandings of behaviors for natural and human system dynamics. [17]

And thus we find people who prefer death—those who have lost much or have nothing more to lose, people who are sick, those who are old, those who are too poor, people who are heartbroken, those who have broken spirits, those who are in pain, people who cannot bear what they know, those who are hopelessly addicted, those who are alone, people who are destitute, those who cannot live with hurt, those who hate and destroy others, people who have destroyed themselves.

So we also find people who prefer life: people who have recovered, those who are renewed, those who have enough, people who find hope, those who have faith, those who overcome pain, people who are wise, those who have no further needs, those who find companionship, people who are generous, those who learn from their mistakes, those who love, people who create.

Those who prefer life must face the same challenges, but they move forward with hope that life will renew their spirits so they may somehow persevere. Isn't it time for terrorists and suicide attackers to "come in from the cold" and modify their conduct to choose life? There's a certain cognitive dissonance that allows people to combine prayer and murder other human beings. That same dissonance allows highly disciplined scientists and soldiers to wage war against our own species. How can issues like this ever be resolved? People will only change when they become aware of the dissonance. [18]

No religion has the ultimate answer unless it shows **all** of us how to live. No economic system has developed any real practice for equitable distribution. Technological development hasn't given viable options for everyone, since it has also been employed to create weapons for destroying part of our own species. Nor do political or economic systems work by excluding people from the natural resources the world has to offer. No system should have a monopoly over how people think and feel. We know that nearly all social systems fall short of their goals when people are manipulative and act selfishly. In our hearts and minds, we know that the social systems will fail unless they provide a focus on life for everyone.

To answer a previous question: *Improving is choosing life, although there's also creative destruction.*

51.

All that exists, all that is animate and inanimate, owes its existence to the Source, which helps to nurture, sustain, protect, motivate, feed and cover under its protection. All these systems harmonize spontaneously.

6.51—Questioning the source

The source of all we know that exists is our world, and on a larger scale, the universe. Everything that is animate comes from what is inanimate. Matter is interrelated to energy, as space is to time. These relations arise spontaneously among all systems. There's no need for human oversight of natural processes, and any attempts to artificially manipulate natural systems usually have undesirable outcomes. [19]

The laws of the universe have nothing to do with beliefs or social systems. Living ecosystems have developed over millions of years, independent of human contact or intervention until recent history. Although we have been around a relatively short period of time in geological terms, we've assured ourselves that we're the main protagonists in the story of our planet, mainly because we're the only ones who are writing the story.

As the heroes of our story, we expect triumph over nature instead of learning from it. In consequence, we fight amongst one another when it would have been better to share. We discover difficult ways to do things, instead of looking for easier ones. We compete for more of the best, even if it means leaving others with nothing. So by becoming selfish, we never learn what would have been better for us and are somehow proud of it.

52.

The Mother of all things is our common origin. Knowing our Mother, we will come to know her other children—how all cycles open and close, begin and end. We can discover an inner vision of even her smallest system—through the eyes of a child.

6.52—Global impacts

It is interesting that we want to dominate our natural world and enjoy all her benefits, without any guilt or consequences, no matter how destructive, selfish, or violent we may be. If Gaia were a woman, we would be accused of rape, for we have certainly violated our trust as the privileged species who inhabits this world, but have ravished her for personal pleasure without giving anything back.

Michael McCurley

People at times make empty arguments to justify the unjustifiable. So we hear things from sex offenders like: "She wanted it." "She's just a slut." "She deserved what she got." "She tempted me." Such excuses are used to cover reprehensible behaviors for assault, sexual chauvinism, and abuse. Sexual offenders, even in our permissive societies, are dealt with harshly. Why are we not more considerate of what we do to the world we live in?

Let's examine whether or not we have anything to do with this planet we inhabit.

If we cannot treat our own species with equitable respect and consideration, what can be expected for the rest of the species who share this world with us?

Consider for a moment that men at different times are intentionally creative or destructive. Enthusiasm varies from moment to moment for each. People produce entropy and contamination from all of their activities, whether they are intended or not. Mental and physical contamination from those activities accumulates in our environment, which is inhabited by all the species of this planet.

In previous centuries, people were limited to exploiting their immediate environment on the habitable land masses of our planet, which comprised about 30 percent of its surface area. Forests were exploited for timber and fuel in wood based energy economies, before coal became a substitute for wood as forests and readily available supplies of wood became scarcer. [20]

Burning wood, and then coal, as raw materials from forests ran out, and more recently petroleum derivatives and natural gas, have all produced large quantities of gaseous by-products which are greater in volume and weight than the fossil fuels we burn, since these combine with oxygen when they are burned to produce carbon dioxide (principally) and other byproducts. All of this has been added to our atmosphere for hundreds of years. Although part of what we produce is reabsorbed, the proportional mixture of gases in our atmosphere is changing. It's also notable that a portion of the carbon dioxide we expected to turn up in our atmosphere has turned up in our planet's oceans instead, and these are becoming more acidic. [21]

Why is this important?

The sun is the primary energy source for this planet, not fossil fuels. The average worldwide median temperature for the atmosphere, depends upon a relative mix of gases that retain a certain amount of heat from the sun, allow sunlight to reach plant vegetation for photosynthesis that produces food sources and oxygen for living organisms, and radiates excess heat back into space in ways that prevent the atmosphere from overheating—which would threaten most life forms. What happens if we disturb that balance, or what might also happen through global contamination of the world's oceans? These questions are debatable as long as it is unclear that we are causing the damage. However, it's unlikely we could undo what has already been done, especially if the impact of human activities still increases.

We have been dumping millions of tons of wastes into the atmosphere on a daily basis for the past century, starting with burning wood as our primary fuel hundreds of years ago. Do we really think we can do this without causing any consequences to our climate or environment? While we pretend that what we do has no effect on global systems, because it's convenient to believe that, we're beginning to note measurable annual increases in carbon dioxide and elevated temperatures in locations around the world, while seasonal changes and weather events are beginning to produce unexpected results. [22] And what about our impact on other living species?

Should we be concerned? Climate change skeptics say no, there are no real lasting effects from our activities on the overall global system. Other environmentalists warn that we have already surpassed saturation points for normal balances and elimination of contaminants, which will lead to greater weather and climate events with negative or unpredictable. Consequences. To imagine we aren't changing anything globally is wishful thinking, not scientific fact. Political "expedience" has also blocked meaningful or significant changes to energy legislation because we're addicted to the energy sources we already consume. [23]

In a movie, all of these would be excellent ingredients for a "Perfect Storm." The players are present and the stage is set. All we need is the director's signal for the crises on a larger scale.

The fact we don't see immediate global consequences, doesn't mean we aren't about to see them. Keep in mind that geological time scales are quite different than human ones. Minutes, hours, days, weeks, or years—these are human time frames.

Decades, centuries, and ages, are more in line for geological changes. If we modify global parameters without realizing what we are doing, we will be unable to change them back later, even when we want to. Once the fossil fuel age ends, there's no return. But that's not all we have to face. [24]

Another question is: *Are we ready for a transition to a renewable energy economy?* We need to take advantage and use some of our fossil fuel resources to leverage the shift to other alternative energy resources before most fossil fuel resources are exhausted. Will we respond in time, or react at the last moment? This seems doable on paper until we try to put it into practice. [25]

For people who like drama with special movie effects, the general consensus seems to be: *we'll respond at the last moment.*

53.

One who possesses the smallest grain of wisdom will follow the Systems Thinking Way (the Tao) and not depart from it. This way is level and straight, though many prefer more tortuous paths. People seem to know much about the ways of the world, but what do they know of the Universe?

6.53—What we follow

What most people follow is self-interest. While this makes sense for most species, it doesn't always make sense as an intelligent, well-considered choice for rational human beings who live and must share resources in a global society.

Self-interest is a fundamental basis for capitalism. Supposedly, if we were all able to freely pursue self-interest to the fullest, all the blessings of our kingdoms would flow abundantly and we would be better off than we are now. While there's a grain of truth to this, it has nothing to do with people following self-interest to the maximum. Uncontrolled self-interest produces a Tragedy of Commons instead. Even a small amount of common sense shows this. We won't have enough if some people are uncontrollably selfish in a limited resource environment, no matter how nice being rich may be for those who get the chance. While a few become rich, a much higher number of people become poor, and too many of these who have suffered from extreme hunger, sickness, and poverty have died. Since this situation has existed for centuries, it's relatively easy to fault victims somehow for their own afflictions. [26]

Self-interest is a simple euphemism for selfishness, which is at the heart of capitalist systems, but never absent in socialist systems either. [27]

Self-interest shouldn't be confused with self-reliance, however, since the second term implies that people can think for themselves. Selfishness is usually blind to everything else. Self-reliance is more practical in terms of awareness.

The System Way, shown to us by the Tao, is simpler than most explanations. It has fewer twists and turns, and is less torturous than other methods. Of course, it doesn't offer easy riches, high profits, or investment returns since it focuses on underlying causes for things which happen around us. Once we understand these, we'll have better ideas for how things work.

People may have extensive knowledge of the ways of the world, but what do they really understand? What they think cannot compare to a greater understanding of the universe.

54.

Cultivated systems have roots which run deep. What you cultivate in yourself remains inside you. If you cultivate a family, it becomes a permanent extension; if you cultivate a community, it lives and grows; if you cultivate a state, it will flower and bloom. If you cultivate a world, it will become universal. And yet you'll find all of this within you.

6.54—*What we cultivate*

What we build or cultivate comes from who we are and how we think. What we build and cultivate also *determines* how we think. This circular or cyclic process may allow us to concentrate on what is immediately important, but it's not so effective in other ways for the same reason—when we concentrate on immediate concerns—we'll miss the consequences we must face at some point later on.

If we cultivate a family, our thoughts, efforts, and beliefs will be focused on that. If we're active in our communities, then our actions and efforts will move in that direction. If we go into politics, we'll direct our thoughts and actions along political action lines of a state. Each of these moves our thinking and actions in different directions, but these are related to outcomes which create exclusive perceptions and ways of thinking. Thus we have different mindsets and paradigms all over the world, but people don't realize that while mindsets may seem quite different, basic human behaviors are still fundamentally the same, and the types of mindsets people have influence their behaviors in divergent ways. [28]

Cultivating the world is the challenge we have that remains. That challenge goes far beyond our normal concepts of family, community, or state. It goes beyond ourselves as individuals, to who we are as a species.

People won't see what they don't recognize. That is why the ideas of some thinkers speak loudly to the people of their times, but not so much when the perceptions of people change. While the ideas of other thinkers, if spoken at all, were spoken of quietly because they were not in harmony with popular opinion. Some ideas have grown stronger and become more clearly understood over time, as perceptions of people change. This is because certain ideas are timeless. These ideas may not always be understood clearly from the beginning, but since they never change over time, they become understood later when we are ready for them. Temporal ideas (and fads) are superseded through experience and time. These are part of the din and the roar—the excitement of the moment. Once they pass, the quiet and unobtrusive ideas are the ones that remain. [29]

A systems thinker doesn't need to leave his house or collect extensive field data to cultivate an inner awareness that goes beyond external perceptions. So instead of focusing on immediate perceptions, which are distractions people are busy cultivating, this thinker develops an awareness to see beyond temporal manifestations into the essential elements of everything else. This universal awareness won't change because it's timeless and waits for the moment when the rest of us will notice it.

Chapter 7.

Governing Causes

太陽

55.

Truth within systems must be constantly approached anew. Understanding the harmony of systems is to have an immutable inner vision. Provoking growth or disruptive manipulation invites collapse and ruin.

7.55—Perceptions and truth

Our perceptions cause us to believe whatever we assume to be true. If it's common knowledge that the world is flat, then anyone who claims it is round would be a nutcase. So we can see that people aren't so much concerned about the truth, as their perceptions which cause them to think as they do. People have died for this reason alone.

Take, for example, our once geocentric point of view. Practically everyone *knew* in the 16th century that the sun orbited the earth. It was common sense. One free thinking Italian Dominican Friar named Giordano Bruno described a cosmology in which the stars were suns, not unlike our own sun, and that contrary to popular belief, the earth orbited the sun. Since he pointed this out, among other things, to those who might listen, he was prosecuted by the Inquisition for disseminating these and other beliefs, and then burned at the stake as a heretic. [1]

It takes more than what people think to determine the actual truth about any situation. This is important for approaching the study of new problems or systems, because we may interpret counterintuitive signs differently than what they actually are. Once we get the wrong idea about a system, it's extremely difficult to undo our thinking and change it to think in any other way. The fact is: *part of what we think and believe is inevitably false*.

People don't like being wrong about what they think, and will often go to great lengths not to change their minds, even when it becomes evident they are mistaken. So in comparing perception with the truth, be prepared for people to be inclined to go with a perception before the truth, except if that perception is true anyway. [2]

A person who studies a system should approach it with an open mind, and be willing to constantly find truth again—as if discovering it for the first time—even whether the original perception was certain or not. In this case, a perception may be a disadvantage unless it has some relation to the truth. Because of this, it's necessary to have an inner vision that's independent of external influences. In this way, we can learn something new, instead of depending upon suppositions we pick up. [3]

A person who works with a system needs to avoid the temptation to manipulate it. This is difficult to avoid because people often force or manipulate systems out of selfishness, trying to make them grow, violating or changing their parameters without real understanding. Those who do this will most probably fail and may ultimately destroy what they are attempting to do.

56.

One who knows doesn't have to say what one who talks doesn't know. The entire world must be seen as a single dynamic system.

7.56—Perception and belief

People tend to predispose arguments to their own points of view. Since they naturally believe whatever they assume to be true, this lumps truth and belief together in ways that are difficult to distinguish or separate. It also tends to mix facts with fiction and fallacies, which are somewhat easier to distinguish, but not if people are partial to what they understand.

What's presented as scientific reasoning is also sometimes flawed because of this. So it's easier for people to mislead themselves than to be misled. People are often comfortable with what they believe or understand, not because it's true, but because it fits within a paradigm which is attractive and easy to assimilate. This also explains paradoxically why knowledge from apparently objective methods of science is somehow diverted for doing things such as killing people or creating weapons of mass destruction for purposes that are quite subjective.

Leave out, then, the obvious contradictions, counterintuitive paradoxes, or inconvenient truths. Education helps to cement a patchwork of apparent half-truths together, along with generous help from the media, public tax systems, and corporate sponsors. We learn, or we're conditioned by what we learn.

There is no end to this story, and no moral other than we deserve whatever we believe because we believe it, regardless of whether it's true or not—because if it isn't true, it ought to be, and isn't that almost the same thing?

Somehow a few people still have a sneaking suspicion that something isn't quite right, but they're not encouraged to express dissension or contradictory opinions because of what others might think, unless, of course, they have joined one party of thought or another, where most of their thinking is already done for them. Although people are "officially" encouraged to think for themselves, they're not actually motivated to do this— either they are invited not to interrupt in classes, or in other venues they're told not to waste valuable time unless they are getting paid—or will pay for it. That's what good "business" is all about.

Instead, people are quietly invited to conform to whatever *Groupthink* has provided for them in two or three basic flavors. These aren't very different from one another. Since the underlying basis for those flavors is the same, they only give people the illusion of choice.

With the advent of the Internet, it's easier for people to appear to be more intelligent than they really are, and appearance or illusion is becoming more important than what is real, especially as virtual reality becomes a major component of people's lives. What *is* real?

Once reality becomes questionable, anything is possible, while the value of truth becomes irrelevant.

When one man has reduced a fact of the imagination to be a fact to his understanding, I foresee that all men will at length establish their lives on that basis.—Henry David Thoreau [4]

The real world that sages understand is not so ambiguous. It doesn;t have to be changed to impact peoples' lives. It isn't necessary to speak of the world to the wise. What we need is to learn how to live wisely. The real nature of life is still a mystery, but it's not something we can easily destroy like other things we are responsible for (because in doing so we would also destroy ourselves). The entire world is one and we only need to realize that we're a part of it, without the need to dominate it.

57.

One can learn the secret for gaining the world by letting it be. The more we try to change things, the more they stay the same—or get worse. People transform themselves. My help to change them is unnecessary.

7.57—Let it be

People have become so manipulative of this world and one another, that they cannot imagine being any other way. This utter co-dependence is a paradox that often brings out the worst in people instead of the best.

Why are the weak, the poor, and the meek called blessed, and those who control the governments and riches of our world's resources most often damned?— because we would have been better off without them. The world works best by letting things be, and we've done pretty much the opposite of whatever that would be—to the point that we have no idea how we could live, other than the ways we do.

System Dynamics may also be selfishly employed to help corporations increase profit margins, without tackling more important human problems. People have become so obsessed with economic growth that they cannot imagine what life would be without it, but they now have to face declining business structures, aging populations, scarcities of resources, environmental degradation, destructive competition, terrorism, and wars. This has produced a great divide between what people come to expect and what they actually see or experience. [5]

Added to this is the fact that many of our world's "experts" often suggest we act in ways which have the opposite effects for what needs to be done for most of us. These "experts" insist upon growth as the solution for economic systems with converging symptoms of decline. As people apply their "solutions" with increasing force or insistence, the consequences for the systems they've tried to change naturally grow worse. Although it seems unkind to say this about people who act with good intentions, it would be better for everyone if these people had never acted at all. For converging systems with diminishing returns, such efforts are wasted. [6]

In most cases, self-regulating systems don't require human action and tend to resist intervention. Results from trying to change these systems are often unexpected or counterproductive, and rarely match the expectations of those who try to change them. This may not be good news, even for students who are careful about studying System Dynamics. There's no quick fix in Systems Thinking, no magic bullet or formula that suddenly makes our problems go away and leads us to that utopic paradise some people have promised. That's because we've done too much damage in other ways already, and what we have done for centuries cannot be easily reverted or undone in a few moments, no matter how good our ideas or intentions may be. People cannot change how they think unless they're aware and motivated to do so, but most aren't eager to change what they learned, especially if they've been conditioned to be selfish and have mindsets which are easier to maintain than to do otherwise. [7]

A simpler world, in which everyone has enough, would require fewer resources than the one we live in already. Those with nothing now would have nothing to lose if we tried to make our world more equitable. Since people with power and control have everything to lose, they will do whatever is possible to keep things as they are—to maintain the status quo. It makes no difference who has less, as long as those with more can keep it that way. This doesn't take into account, however, the impacts of cumulative damages over time.

So people who are contentious and nervous about this state of affairs in the world are now divided into two groups:

1) Those who have the most to lose if they must give up part of what they have (since they control most everything already).

2) Those who have almost nothing and must struggle day to day in order to survive, and have nothing to lose if they try to recover what has been taken from them.

The world may adjust itself eventually (perhaps sooner than people think) without any intervention, even as the overseers in the corridors of power do their utmost to maintain perpetual supremacy. The rest of us will be observers of these events which are about to unfold.

The writing is on the wall. There isn't much doubt about which group could eventually prevail. That worries people who have much to lose and gives hope for those with much to gain.

58.

When a government is discrete, people are simple and happy. When a government is repressive, people are astute and unhappy. What could be a definite purpose or end to this process? A wise one squares things without cutting, sculpts without disfiguring, straightens without forcing, and clarifies without disillusion.

7.58—Aims of government

The goal of nearly any government is to control its subjects—to make people follow a political-economic system, which often ends up being the flavor of the month, one that most people don't like much anyway. So the goal of those who are subjects of a government is to resist control. This is the natural behavior of human beings in any political system, and it's irrelevant what the politics of that system might be. It is unusual that many people never notice this. Otherwise, citizens and their governments might be quite different.

For worst case scenarios (without citing innumerable examples) repressive governments of any flavor, (democracies, federations, oligarchies, dictatorships, socialist states, and republics) try to force people into conformity, but eventually force themselves out of existence. Overly lenient governments govern through inaction, but eventually succumb to the corruption of selfish civil servants who end up serving their own interests. Then these regimes collapse or fall into the hands of more repressive forces that promise to reestablish law and order through whatever condiment of nationalism that is available in that place and time. And so on.

For best case scenarios, there's a semblance of choice, often code-named "democracy," and lip service is given to personal freedoms or whatever appropriate rhetoric people have become accustomed to. These governments rule less overbearingly, but aren't (yet) too corrupt, and people who have more freedom to be themselves recognize the value in leaving what is well enough alone. Still, with nothing better to do, political parties vie for their turns in the sun through political elections, as the political pendulum swings back and forth between the official or the opposition parties. Those who become bored or dissatisfied, notice that nothing really improves. Some change their minds from one election cycle to the next, but never notice how their country's political system oscillates like a pendulum as a whole. [8]

This suggests implicitly, as many say, that the best government is one that rules little, or not at all (although it's somewhat of a paradox that the political parties which give lip service to this point of view are often more militaristic and repressive of human freedoms than others). But to follow our proposition to its ultimate consequence would imply we could live in a political vacuum, which has never been the case. Most likely, that vacuum would be filled with something different or worse, if nothing were there in the first place. For those who cannot imagine what that might be, think of the mafia, drug cartels, or other well established, but less reputable social institutions. Or consider the influence of an external power like China. So the case people make for anarchism or social utopia wouldn't work—that is to say: *we don't trust ourselves (or others) enough to live without government*—though wouldn't it be nice if we could? [9]

59.

The virtue of knowing systemics is to discover a possible erroneous path before it ever occurs. This invisible path to improvement is possible through the Mother of Strategy, the secret to long life of a system, and a durable vision of its future potential.

7.59—The invisible path of improvement

The problem with a path for improvement is that it's invisible. There is no way to compare it with more direct productive results. So when improvement programs are compared with productive results for time management purposes, there's no question about which will be eliminated. This shows that in spite of our best intentions, even with excellent results, most programs for improvements are tried but then eliminated before they can be implemented for long term results. [10] (How could we compare a tangible result with the intangible benefit of eliminating a possible problem that doesn't exist?) Then, the fact that the improvement program hasn't produced tangible long term results is given as the reason for eliminating it. This use of circular reasoning isn't recognized. Instead, other cases are cited which show similar conclusions as further proof that improvement programs don't work effectively on a long term basis. This absurd determination of self-fulfilling prophecy would be funny if it wasn't taken so seriously, but we find this repeatedly applied in corporate environments without any awareness that it's happening. [11]

So the disadvantage of eliminating problems before they ever occur is that you'll never be recognized for problems that don't exist. Police forces would never be recognized for fighting criminals who aren't there. Social organizations would be unnecessary if there were no social problems to combat in the first place. Technicians would not be required if there were no problems to fix. People work and thrive right now with problems. Eliminating them would not be as impressive or as easy as it might appear, since people wouldn't be pleased if the problems (or enemies) we know so well disappeared, unless we also change how the structures of society would work without the problems we're accustomed to.

Ideally, we should identify potential problems before they occur and avoid them, but our societies are not structured to work that way. We are taught instead to face and fight problems until we have defeated them as the enemies they are (and should be). What we're not taught is that we often are the ones who create those problems— so in that case, the enemy would be us. Since it's inconvenient to fight against ourselves, we ignore the possibility and look for other causes or demons to blame in others (the Mother of all wars), not realizing that simple recognition of mistakes would give us the means to correct and solve our own problems without any more complications than admitting that we're sometimes wrong (which we know anyway). People who are defeated in war will certainly discover this, which can be useful for recovery. The victors face the paradox of having won, but re-vindicate what's often wrong.

Sometimes people don't reason well because there's no reason for them to do so, or they may find reasons not to do so.

60.

Systems demons will have no power to harm us *if* we learn how to avoid them. If only a government and its people refrain from doing mutual damage, all other benefits of that kingdom will come forth.

7.60—System demons

At times, when problems appear and system behaviors seem out of control, people might say, "*There are demons in that system!*" In a sense, they would be right. Those who have no understanding or control over systems may have reasons to be afraid.

But people are usually more responsible for system behaviors than demons are, even if they refuse to be accountable.

Those who are wiser know how to avoid certain behaviors which are sure to cause trouble. And by knowing how systems work, they can help others troubleshoot and resolve problems before these produce serious consequences. [12]

Much of this comes from simple common sense related to systems. But people often choose difficult and roundabout ways of doing things.

Consider a government and a rebellious group of people.

They could fight, of course, to see which side obtains some advantage over the other.

Or…they could negotiate to avoid doing mutual damage to one another (which would only escalate), so that neither side would be damaged. What benefit does such a conflict have for humanity? None at all.

So if both sides come to an agreement, everyone benefits overall. Without an agreement, both sides lose what they otherwise might have gained.

61.

A great kingdom is like a valley where rivers converge, where all that exists flows within a feminine current that conquers through quietude. When a larger system is inclined to provide energy to a smaller one, both will mutually benefit.

7.61—Interacting Systems

Large systems will eventually produce converging behaviors. That is, as they grow and interact with other systems, the growth rate will decrease, converge, or stop at a certain point, which some refer to as a tipping point. Systems with explosive growth reach tipping points sooner than those with slower growth. Such systems may also overshoot and collapse. [13] [14]

As large and small systems interact, their relations are often complementary. A larger system may lend some of its energy or resources, for example, to a smaller one. The smaller system may serve as a buffer, or open a mutually beneficial relation with a larger one.

This works on many levels, from personal relationships between people to international relationships between countries. The same basic principles apply, regardless of the size, scale, or time frames of the systems involved.

There are masculine and feminine currents (Yin and Yang) which interact in all animate and inanimate systems. We will be able to respond to the complementary behaviors of those systems, as long as we understand them.

62.

An understanding of systems is a universal deposit of all things. Those who safeguard this knowledge of systems can check one another's work for errors. A good model is an unmatched gift for those who otherwise might stumble.

7.62—Awareness and understanding

True awareness and understanding can be a greater treasure than all the trappings of wealth and power. This is because understanding is more important than the external physical manifestations we observe. To be aware of causes for what is manifested also gives us the potential to create almost anything.

Why is this so?

The resources we need to create our future already exist in one form or another. Knowing the root causes for what appears and is manifested, shows us how to create new possibilities. [15]

So it's better to have wisdom, awareness, and understanding than to have physical wealth, because while one may last, the other will not. Watch what happens over time and you'll notice that what you have one moment is soon gone the next.

And you can be aware of all this without moving or making any effort to know it.

63.

An understanding of potential inertia can produce great results, especially if we understand the growth of self-organizing, living systems. All we need to do is pay attention to small beginnings. From those beginnings, we will eventually see great results.

7.63—Living systems as models

We would do well to study how living systems are organized and develop. Living organisms reproduce from extremely minute amounts of living matter. Just a few simple cells, for example, are codified with everything needed to guide the development of fully grown organisms. But defining and modeling life isn't that simple. [16]

Human systems which mimic living ones are often successful. That doesn't mean, however, we should play very much with living systems-such as genetics. Our species is still relatively immature (based on our behaviors in the last century) to experiment much with biological systems, given mixed results:

—weapons for biological warfare

and

—biomedical advances which have improved health.

[17]

Unfortunately, we've demonstrated significant advances in opposite directions. Some of them, we would be better off without. Others divert our use of resources for less than beneficial purposes. But finally, we've made a few advances which have contributed to positive results for our species.

People who easily make promises aren't likely to succeed, because they make the mistake of assuming something is initially easy, when it isn't.

Those who are most likely to succeed start out by recognizing the difficulties of what they want to do, and then find it easier to accomplish when they persist until they achieve it.

Chapter 8.

Control

天地

64.

If the basic aspects of systems are understood, we may eliminate certain problems before they grow beyond our control. And although system processes may seem beyond us, any journey starts from where we are standing—with the first step. A wise one learns objectively to induce others to learn as well.

8.64—Awakenings—how things start

The greatest journey begins from where you're standing. Take one step in the right direction, then take the next. That's all. Then you can progress step by step until you accomplish your dreams. Everything that begins and ends has a tipping point—the dynamic point at which something starts, and an eventual point at which growth or development stops.

Life starts with growth, and so does whatever we creatively imagine. An idea germinates in our minds, a small inkling or desire of what could be—a beginning.

Before that there was nothing. But then, anything is possible! What is the motive that makes what we create come into existence? Hope. What enables us to understand the world? Our perceptions. What is the basis of our existence? Life. Without ideas, hope, perceptions and life, who we are and what we've done wouldn't exist.

A living human organism starts with the fusion of two gametes (sperm and ovum) into a single zygote cell. This fusion takes places in about a second. And yet all the complexity of a complete multicellular organism, a human being, unfolds from that cell. A fully developed adult may have 70 to 100 trillion cells. Yet all of these are part of the same person derived from that single zygote cell. [1] How different or similar is one person from another? Although people appear to be outwardly different, 99.9 percent are genetically identical if compared between the same sexes—regardless of geographical locations—and 98.5 percent are genetically identical if compared between different genders (male and female). [2]

How does a chicken come from an egg? At times, there are mutual paradoxes that defy common sense. The same is true of contradictory human behaviors. So we have reached equal points for great opportunities and dangers. It's not entirely clear what our future course might be, but that's part of the adventure.

Certainly it does no good to warn that we have embarked on a path for which we're doomed, so we would abandon all hope. But it would be equally unwise for us to pursue our interests and ignore the consequences of our actions. Feedbacks tend to amplify their effects over time. What starts out infinitesimally small can become exponentially enormous in terms of positive or negative results, and those feedbacks can tip or change what we know into something else. [3]

65.

In ancient times, systems thinkers of the Tao, who were versed in its arts, didn't attempt to instruct people, but only learned to maintain a state of simplicity. Why are people so difficult to govern?—because they're too smart! They have lost sight of the norms and means for Systems Thinking.

8.65—Criticizing governments and belief systems

It is almost impossible to talk about governments without being severely tempted to criticize them. Most governments, at one point or another, have trampled upon the rights of the people they rule. In extreme moments, they've murdered their heroes, suppressed their populations, eulogized tyrants, massacred the opposition, and usurped or corrupted whatever authority was entrusted to them.

But patriotic history forgets our darker moments and commemorates brave deeds against evil foes who, on one side or the other, are only ourselves. We haven't fought against aliens, evil trolls, giants, or monsters. Instead, we've fought against different cultures, languages, politics, and religions of people who are very much like ourselves. Somehow people became victims or aggressors of these battles because they were conscripted into armies to fight and serve opposing countries and ideals, or they were collateral casualties of war. [4]

What would external observers see of these wars we wage? They would only see us fighting among ourselves.

For external observers, there are no borders, boundaries, countries, economic or political systems, religions, or philosophies. There are only people—how people behave and interact, and what people have created or done to this planet and one another. So what most people follow within the realm of their personal paradigms or belief systems would be irrelevant.

Some individuals have attempted to use intelligence to dominate the political systems of their countries. This has worked with various degrees of failure, to use another term instead of success. However, a few who haven't acted in direct politics at all, like Mahatma Gandhi, have conquered entire empires. Somehow when a system functions in ways that are fundamentally flawed or wrong, it may not take more than a single person who understands the true nature of the confrontation to bring about its decline. [5]

Although this seems entirely fictional or apocryphal, strangely enough, it's true.

66.

Why is the sea the king of all rivers?—because it's **beneath** them. A wise one instructs people with humility to facilitate their learning, so no one will have anything to fight against.

8.66—The basis for life

The kings of all rivers are the oceans, which joined together with the atmosphere, surround the circumference of our planet. This circulatory system of our world provides the basis for all life.

It's interesting to note that hydrogen is also a primary element of the stars, which is sublimated with oxygen to form water—the supporting matrix on earth for living organisms. Without water, life as we know it wouldn't exist. So besides the gravitational analogy that oceans are the kings of all rivers, two reactive components, hydrogen and oxygen form water, a neutral medium that sustains life.

This suggests how leaders could govern the people of our own species if they were wise enough to do so. Aggressive domination isn't necessary to rule. There's no need for reactive egotism in our relations with other people. Leaders will meet with uncommon success if they are humble with the people they interact with and govern. [6]

In this way, they could lead without resentments or conflicts because people would have no reason to fight against them. And thus they could tackle systemic problems for better government without engaging in emotional conflicts that completely distract many leaders today.

67.

Three treasured principles of systems thinkers are above all others. These are: be compassionate, be moderate, and avoid trying to be first in the world. One who is compassionate is brave; one who is moderate is generous; and one who first follows others is fit to lead.

8.67—Governing strategies and paradoxes

On the surface, any attempt to employ a strategy of using compassion, sobriety, and not trying to be first in the world, seems to border on insanity if we consider how these principles could be employed to influence and rule a state. At first, we may be hard pressed to find any examples that would show this—and then we might remember.

This has been done many times before, and successfully too!

"What?" you might exclaim. *"People who are compassionate, sober and humble, certainly seem too naïve to have any influence or impact upon the government of any country."*

But if we look closely and carefully, we can find superlative examples of these leadership principles. It's just that they won't leap to our attention. The people who embodied these principles haven't tried to call any attention to *themselves*. What they wanted to achieve was more important.

Let us look at who a few of these people are, and what they have done:

1) Henry David Thoreau—wrote quietly and spoke out against slavery, unjust wars, and unreasonable government. He helped spark a movement that paradoxically led to the American Civil War and the emancipation of slaves in the United States. [7]

2) Mahatma Gandhi—led India to freedom by using passive resistance to defy the British Empire and helped his countrymen to found an independent government for India. [8]

3) Martin Luther King—led a protest movement for black Americans that led to greater equal rights for his countrymen in the United States. [9]

4) Nelson Mandela—spent the greater part of his life in South African prisons, but later was released to become the President of South Africa and end a repressive system of Apartheid between whites and blacks. Some difference! [10]

5) Aung San Suu Kyi—started a civil and political movement in her country and was elected President of

Burma (now called Myanmar), but she was rewarded by spending more than 20 years isolated and under house arrest. In spite of that, she has been an inspiration to her people to this day. And like Nelson Mandela, "The Lady" is now considered worthy of respect, even by leaders of the government that imprisoned her. [11]

6) Mikhail Gorbachev—introduced "glasnost" (openness) and "perestroika" (restructuring) in the Soviet Union (U.S.S.R), leading to the democratization of the communist party, which ultimately transformed the communist Union of Soviet Socialist Republics into a Russian Federation of Republics. [12]

7) Deng Xiaoping—succeeded in transforming the economic system of communist China, by bringing rapid sustained development of that system without leading it to collapse, as had happened to the Soviet Union. Although this was accomplished with certain undesirable side effects, it's surprising it could have been accomplished at all in a system containing more than a billion people. [13]

When our thousands of Chinese students abroad return home, you will see how China will transform itself. —Deng Xiaoping [14]

Did these great individuals act alone? No, of course not. What they did wouldn't have made sense or have had any effect if others didn't follow them. But these true leaders inspired millions (and interestingly enough, one another) by following principles that others surely thought would fail—and yet they succeeded. Although we may not notice, people in our world are quietly doing similar things. These people develop abilities to transform the world, not for themselves, but for everyone, one person at a time. While they may not have always succeeded in the ways they envisioned, the fact they accomplished what they did, has produced uncommon successes that have impacted the world we live in today.

For those who express serious doubts, keep in mind that what these people did actually happen and can be confirmed. We can develop those same capabilities and find our own examples if we want to.

68.

A good warrior is neither aggressive nor irritable. The best way to beat a competitor is not to confront him. The best way to employ someone is to serve under his orders. There's virtue in not struggling against a system.

8.68—The art of confrontation

Confront your competitor by not confronting him. Rather than choosing contentious issues as sources of conflict, be the first to resolve what you can on your own. If you want to be free of your enemy, first you will need to free yourself from issues which are the sources of greater problems or conflicts. Choose your own ground and then step back. [15]

"Violence," said the writer Isaac Asimov, "is the last refuge of the incompetent." Unfortunately, both violence and incompetence are still much in vogue these days. (quote from Salvador Hardin in the ***Foundation Triology***, by Isaac Asimov) [16]

Remember the enemy within. You have better things to do than to fight against yourself. This is basically what we do when we fight with others. Revenge and retribution are both tied up in the ego, which is not who we really are.

Passive resistance needs no more from us than to wait until what's right emerges from what has been wrong. It's better to do nothing at all than to act on the wrong side of doing right. Passive resistance is a powerful form of social inertia.

What people compete for would be enough if some people didn't try to take everything for themselves. We have better things to do as a species than to compete with one another.

Systems tend to balance with one another over time, so even confrontations tend to resolve themselves in the long run.

69.

Strategists in the art of war have a saying: *Don't be the first to take the initiative.* Thus you advance without moving. There's nothing worse than underestimating a competitor.

8.69—The art of war

Would your enemy fight if he had no one to fight against? Every conflict needs two contenders. The playing field remains empty if no one else plays. Convention tells us that we need teams to play lethal games of warfare. Wars wouldn't exist if people didn't play. [17]

Never give your enemy a reason to attack or a target to encounter.

People die soon enough. There's no need for us to hasten the process. To the victor goes the spoils and responsibilities for unexpected consequences. War is never good business for long, except for vultures and undertakers. What begins as one form of victory, usually ends as another form of defeat. It makes no sense to destroy an enemy if you're also destroying yourself.

Even terrorists should suspect that their "enemies" aren't unlike themselves, and eventually they should stop fighting or die. But for that, they would need to find better things to do with their lives and live. Those who destroy others, quickly destroy themselves. They have no need for our help to do this, but perhaps they need our help not to.

Although vengeance and war play well in television series and in movies, they make no sense within the ranks of humanity, and on a larger scale it makes no sense to wage war against nature or life itself.

Our species will move from one crisis and war to the next until we learn which of the two is unnecessary.

Defeat your enemy by making him your friend. Then he will no longer attack you. In the 18th century, Benjamin Franklin discovered that by making an enemy his friend (through asking him a favor), that person stopped attacking him. This is better than turning friends into enemies, which we do often enough. The less we have to say about fighting, the less likely we are to fight, until we have nothing left to say about the subject at all. In any case, if we were to follow this idea to its logical conclusion, we would have fewer enemies and more friends. [18]

70.

Principles of systems are easy to understand but difficult to put into practice. And yet such principles have antecedents and proven performance. It is a noble endeavor for the few who follow them.

8.70—*Easier said than done*

Most of us are aware of what we should do, but one thing is to say this, and another to do it.

Here's a simple *"To Do"* list we find difficult to put into practice:

1) Become globally aware as individuals and as a species.

2) Care for and conserve what we have for our children's grand children.

3) Develop renewable resources which can be shared now and for future generations.

4) Keep in mind that people and resources are two different things. Using people as resources or things, defeats the value of our existence.

5) Learn to accept mistakes, find root causes, and discover solutions or improvements whenever possible. By understanding systems, we will see how things work and discover how to fix them.

6) Practice real democracy through developing more equitable social, economic, and political actions with participation for everyone.

7) Recognize that we never truly own what we have, especially since we actually borrow what we need to live on this planet. Life is a gift, best shared with others.

8) Respect Nature as the source of all life and resources, including our own.

9) Understand when enough is enough in terms of what we need or can be used. Excessive growth will only lead to overshoot and collapse.

These are fairly simple goals when we first see them.

Why are they so difficult for us to achieve?

Some of us are still quite selfish and haven't overcome the enslaving dominance of our own egos. [18]

71.

Discovering that our knowledge is ignorance leads to self-awareness because many consider that their ignorance is knowledge, which is insane. The wise one, who is untouched by this confusion, has the secret to life.

8.71 When knowledge is ignorance

People who have an agenda for knowledge will find that they end up knowing very little. Knowledge has no value in and of itself. What good are millions or billions of words, unless they make sense and can somehow be used? A quick attempt at using an Internet search engine will show us that. Many people decide what they want to know before they learn it. So what utility is there if people filter what they learn? It's hard to be objective if we decide the answers before we hear the questions, if we determine outcomes before we understand their causes, or give findings or conclusions before we know what produced them.

Much of what we expect is the opposite of wisdom. This confounds and surprises us, as well as scientists and scholars. While we expect to be warned of possible detours or alternate routes along the way, we also hope to find hints of how we should proceed. So then, wrong knowledge is a hindrance and right knowledge a blessing, but both seem very much alike.

When knowledge becomes ignorance, those who follow it will cultivate mental wastelands of weeds instead of fruitful gardens. What sprouts will wither and die, or overgrow and do the same. There are times when it's better to know nothing than to apply the knowledge of so-called experts. And yet there are also times when it's evident that we learn from our mistakes.

If we learn for the sake of learning, we're only fooling ourselves. When we learn out of selfishness, it would be better if we never learned at all. And if we build upon more and more knowledge to create or build ever more complicated projects, sooner or later we will court disaster. Although it may seem to be coincidence, we will see that the best planned projects can suddenly disintegrate or explode like the Challenger Space Shuttle, with tragic or devastating consequences. [20]

True wisdom isn't based on complex knowledge, even if experts say otherwise. Simplicity is the key to wisdom and those who reduce things to simple terms are best understood, while others only confuse the issues they present. Wisdom also comes from within the collective human unconsciousness that is latent inside us. There are times when what we know hasn't come from anything we've learned. We were already aware of it. [21]

What we consider important is actually a tiny fraction of the universe. It's no wonder that however much we know is ignorance compared with what remains to be discovered.

72.

When people no longer fear you, it's a sign that you are attaining great power. If you don't interfere in their homes or overburden them, they will flock to your cause. Know yourself.

8.72 Know yourself

This refrain has echoed throughout the ages—it seems so simple and yet is so difficult to do. Why? ——Because there's a dichotomy within us, which is divided between our egos and the persons we really are—also between the creative and destructive aspects of human nature.

Although your ego helps to preserve you as an individual to survive, there are uncontrolled tendencies of the ego, which bring suffering to mankind. This darker side produces negative human traits we know well, such as avarice, cruelty, envy, greed, larceny, revenge, selfishness, violence, and war. In certain circumstances, your ego tends to complete your personality, but in others it competes with your own personality until it becomes difficult to express who you really are. [22]

Who you are doesn't depend on what you have, how important you are, or what you do.

It's no wonder then that human ego is responsible for the worst kinds of dictatorships, for abuses of human rights, and for forcible occupations of territories which have no real owners. What gives people the right to control what isn't theirs to take?

It's easy to confuse ego with who you are.

The human ego easily becomes a tyrant without limits. This happens often enough in real time for us to be aware through what we can observe every day. Ego justifies the unjustifiable because it strictly limits perception to itself—but it's not the real basis of your identity.

The essence of who you are is within you. There's no need for you to have, possess, do, or be more or less than who you already are. The essence of your identity requires none of this. You have nothing to prove other than to be yourself. Anything less subtracts and diminishes you, just as anything more does the same. Ego does this if you allow it to. Your ego can suppress and replace who you are with mere selfish desires that become fixed on empty relics, material possessions, and of course, money. But money has no real value or power except for what we ascribe to it—what we think or say it's worth. Otherwise, it is only paper and shiny pieces of metal with images, letters, and numbers. [23] Think of what people will do for money, and then consider for a moment that it's really nothing.

The power of such things is within our minds—but is worthless outside of them. So this brings us back to the influences of thinking and perception. Your mind can cause you to believe or do almost anything. The results depend on your motivation and purpose.

This isn't limitless, even if it seems to be. There are mental and physical limits to your mind, just like anything else. For that reason, Systems Thinking provides more objective ways for us to model systems and use visual models, which allows others to track or examine how we think. It also allows you to reflect and review how you think yourself.

The Dark Side

There's another side to us we never mention, unless we speak out against others. This is the dark destructive force of human nature, which is never entirely absent from the best of us, and more often present in the worst. And so this deviation is in all of us.

This force shouldn't be underestimated, because while there are noble examples of people who are willing to dedicate themselves or sacrifice their lives for creative or beneficial causes, there are also many people who are actively involved and willing to sacrifice themselves (or others) for destructive purposes.

Thus we have:

Crime, abuse, betrayal, envy, and lust,

Deceit, destruction, genocide, jealousy, and murder,

Fear, greed, hatred, larceny, and oppression,

Rape, robbery, vandalism, vengeance, and warfare.

Certainly, the list could go on. But that would emphasize a side we prefer to ignore rather than deal with, because we recognize the echoes of these same conducts within ourselves.

Although we fear repression and extermination in our own societies, we may allow others to go to extermination camps, rather than go to them ourselves, without realizing that by doing so, we have helped create the horrors we fear most—in both fiction and reality—because we're afraid to lose or die—though that's what we all will do anyway. [24]

To deal with our Dark side, we fight against enemies upon whom we project the reprehensible qualities we loathe and fear, since this absolves us from recognizing them in ourselves. People who do this never see what they are really doing because they deny it. We also may not recognize that this dark side is within everyone.

The Dark force counterbalances the Creative force within us. Along with the desire to create is also a great temptation to destroy. While we may not be able to eradicate one side of our nature without debilitating the other, we may profit by being aware that both are within *us* as well as others. [25]

That same fanaticism—those same obsessions—can also be easily ignited within us. Take a few moments to examine historical events to consider whether this has happened.

This is what has led us to the violence and destruction of all our wars, even among the most advanced or primitive civilizations. Although this primal force has been sublimated many times when we've wanted to, it can be unleashed with as much fury as the nuclear weapons we've created if we allow it to get out of control.

A healthy awareness of this is wiser than negligent denial.

There's no truly evil side we can fight against or eradicate, any more than there is perfection or utopia.

What remains is to choose which direction we want to go—knowing that everything we build or do will eventually perish, and that some people will attempt to destroy what we try to accomplish anyway, sooner than we want or expect.

Wisdom comes from knowing the truth about both sides of our nature—one which is caring and kind, the other which is grasping and selfish. From this we come to realize that there is no perfect social, religious, political, or economic system (or person). There are only social conventions that we've been conditioned to accept and are comfortable with. What we believe and know is based on what we've been exposed to. Our habits and even our prejudices are learned behaviors. Although we have a certain degree of free will, most of our behaviors are formed by what we've experienced or learned, and these can be manipulated.

Understanding who we are comes from within us, one person at a time, until we share a common consciousness that gives us collective insight and maturity. This is not the mindless fear or totalitarian manipulative consciousness that some imagine, but instead it's an awareness that comes as we grow and evolve as a species.

It would be equally evil to identify only our positive potential without recognizing our destructive and selfish side as well. This is part of the individual wisdom of knowing yourself and all of your potentials.

You are the moderator of your own desires, but in a larger sense, you are also a partner in shaping humanity. [26]

洋銀

Chapter 9.

Wisdom

陰陽

73.

One who is brave but fears will perish, but valor without fear is survival. If we overreact to systems, our responses will be controlled by them. One way to conquer is to understand a system without fighting it.

9.73—What will and won't succeed

There is an old adage which says, "Fools rush in." This is wisdom if we follow it. Overly sure and reckless people are often wrong, and frequently lead others to nearly anyplace except where they need to go. It makes sense to consider what we are about to do carefully. As a species, we tend to be overconfident at the beginning of any endeavor, and conveniently forget how painful or difficult it was to accomplish when we finish. Perhaps that's just as well. Otherwise, there are many things we would never attempt to do in the first place. A good number of them, however, should never have been attempted.

There's a difference between being reckless and being fearless. One comes out of courage, the other from carelessness. Those who are careless quickly face failure. But individuals who are brave, flexible, and resilient may succeed. Those who are quick and careless often only want more. People who are more careful learn to be content and understand when they have enough.

The best way to conquer a system is to understand it without fighting or conflict. When we overreact to systems, we're controlled by them. [1] When we find what balances a system, we discover what makes it work. By understanding what makes a system work, we can learn to work with the system rather than against it, and make things better or succeed in what we want.

74.

There are systems laws that no one can violate without consequences. Whether people are aware of them or not, these laws are impartial, even for the practitioners of System Dynamics.

9.74—Final destination

To put it simply, death is the final destination for everything—when energy is depleted, movement stops, heat has dissipated, and life is extinguished.

Before that last stop—what do we really have to lose?

No one—not even the richest person in the world—has gone beyond that final divide and ever returned. So you can't take it with you, and whatever you have accumulated during your lifetime stays—no matter how little or much. Hopefully, you will have contributed something that becomes a part of this world and who we are.

So, is there anything we can leave behind? Of course!

Some leave contributions to wisdom, others help to better society, collectors have created art centers and museums, and a few have built institutions of philanthropy. The value of how we participate has nothing to do with how rich or famous people become, but it does depend upon how we've been motivated. What is valuable, in the end, is what we've shared with others.

We should realize that Gaia has shared gifts from her resources for all of us to use during our lifetimes. But to keep explosive human development in check, natural lifetimes of people are generally limited to less than a hundred years, general consciousness is limited to individual rather than a more united consciousness, and individual selfishness and competitive behaviors tend to slow human growth. There may be a purpose to these feedbacks for the rest of the world. [2]

Even death is a contributing feedback that maintains a certain degree of stability that might otherwise be destroyed if growth or development of our species were too fast or explosively uncontrollable. Recent history also points to behaviors among our nations and societies that indicate we're maturing as a species.

During the 1970's there were fears that the world's human populations would grow out of control. Birth rates, which were significantly higher than death rates, rapidly increased our global population. Simultaneous advances in medicine also reduced death rates in many countries and as a consequence the gap widened between birth and death rates. There were fears that our populations might overshoot, outstrip food and other resources, and then collapse. [3]

As we reach the crest of the wave in the 21st century, we are beginning to see populations in some countries are no longer increasing. This isn't because death rates in most countries have risen. Instead, birth rates almost everywhere are falling. Women are pursuing professional careers and having children later in life. Couples who get married are having fewer children. So birth rates worldwide are slowly dropping. Awareness and actions for humanity are changing. [4]

What does this mean for the population of humanity as a whole? Global population may eventually stabilize. This doesn't mean our overall population will stay the same, but it may reach a certain level and then vary somewhat over time. When the birth rate is roughly equal to the death rate, the number of people who die is almost the same as the number of those who are born. This may contrast sharply with expectations that population growth would outstrip resources limits, though that may still happen as some resources become scarce (such as water), which may still have great impacts on populations in different regions.[5]

Although there were moments in the last century when our entire species was perched on the brink of extinction, it's notable that even leaders with more limited visions at the time made choices for life instead of annihilation. Learning to make the right choices demonstrates the resilience and flexibility we need to survive and thrive. This is more important than material prosperity. We've always had what we potentially need anyway.

We should keep in mind that we haven't always made the right choices, however, before we blindly follow novel solutions which may lead us further from where we really want to go. There are indications that we're learning, though our rate of progress as a species has been slow and sometimes painful. Our efforts are divergent because some of us move in opposite directions, which has led us to setbacks or magnificent failures. Even these, however, are part of an overall learning process.

An experiment that went wrong

At the end of the 19th century, a new social science called **Eugenics** caught people's attention from around the world. The idea of perfecting humanity was being seriously considered as Charles Darwin introduced *The Origin of the Species*, and his half-cousin Francis Galton wrote *Hereditary Genius*. Ironically, quite the opposite was about to happen. How would human perfection be possible if inferior members were mixed in our populations, who were mentally, physically, or racially unfit? [6] [7]

The idea of eugenics proposed that undesirable population traits could be gradually eliminated, over time, so a superior species would genetically emerge.

What people didn't consider carefully was how those "undesirable" elements of humanity would eventually be eliminated. Efforts to begin doing this were limited at first, but gradually these were implemented in England, Europe, and the United States, and then were taken to extremes by Nazi Germany. This finally culminated with forced sterilizations, executions, and genocide for all people who were considered to be undesirables or unfit. [8]

It was not until millions were killed that people came to realize that far from being humane or scientific, eugenics was a pseudoscience that lacked a genuine basis and had the opposite effect of what people had hoped or expected. Eugenics became an experiment that went terribly wrong. Although nearly everyone now realizes this, many people (including Nobel Laureates) were convinced in the early 20th century that we were on the brink of perfection, rather than the nightmare of a human holocaust. [9]

75.

When a system taxes its people excessively, they may die of hunger. Why are people so difficult to govern?— Because those above them only look after their own interests. People in that society have better things to do and instinctively know that life like that isn't worth living.

9.75—How governments fail

People know intuitively when a government is not for real. After the hype of an inauguration ceremony, are leaders serious about making a better government, or will they just govern for their own benefit? Do these leaders govern using objective principles, or do they govern out of subjective interests? For anyone, personal temptation is high and is easily justified for those who govern within any group or party. Why is it that legislators reward themselves before others (even in poor countries) and essentially ignore the people who elect or support them?

Governments don't give back any more than we give them in the first place—but they often take more than we give. And if we're honest with ourselves, don't believe in magic, and understand entropy and basic laws of thermodynamics (sorry, there are no perpetual motion machines)—we can only expect somewhat less from government than what we started with—unless we find better ways to renew or revitalize the energy and human resources we work with. Interactions between opposing government parties may also escalate through conflicts. [10]

Fortunately, for governments which maintain the status quo within national systems, a majority of people have relatively short attention spans and don't have the patience to focus on government activities for years at a time. So most people who become dissatisfied generally focus from one election cycle to the next, and don't analyze very deeply that not one of a long succession of governments ever fulfills the true expectations of their campaign promises in the long run. Otherwise, most of our problems would have been largely solved by now. [11]

The only explanation for this is that none of our governments have fulfilled their ultimate purpose in ways that all people really need. And this is how governments fail.

Originally the heading for this commentary section was intended to be: *Why governments fail.* But it quickly became evident that overwhelming commentaries about why governments and institutions fail would require an entire encyclopedia.

However, to summarize briefly:

Governments fail when they create expectations for change among adherents to gain popularity, but these quickly escalate beyond reason, and the promise for change never materializes in the long run, even when certain important steps are initiated. For some reason, nothing lasting is ever implemented, except for a few better known practices, borrowed perhaps from another society, someplace else.

When real revolutionary changes begin, they are blocked, distorted, overturned, or thwarted with unfortunate consequences. This has happened because past revolutionary wars haven't had much more success than wars of conquest, but this shouldnt be altogether surprising. And why not? A revolution is intended to conquer the territory of people's minds—if not actual physical territory. An effective revolution wouldn't thwart the minds of individuals by controlling their thoughts. Instead, it would result in a mutual accord among people by choice, not by obligation. For the powers that rule, that would seem dangerous, but such a revolution may be necessary to create a sustainable future for everyone. [12]

76.

A living organism is soft and flexible. A corpse becomes hard and rigid. In Systems Thinking, "hard and rigid" are companions of death. The bigger things are, the harder they fall. For that reason, humble principles of systems can be more powerful than a mighty army. And because of that, mighty armies have fallen for simple reasons that leaders often forget.

9.76—Flexibility and resilience

Two qualities that contribute to the longevity of our species—in spite of our mistakes—are flexibility and resilience.

This doesn't mean that all of us are flexible and resilient, but enough are to allow us to adapt to new situations and circumstances. This helps us to survive—perhaps even thrive—if we use our opportunities to do so.

Those of us who are flexible can adapt to new circumstances, accept and correct mistakes, and adjust to new situations without great difficulties. Such people do well, even when they face adversities, or perhaps because of them...

People who are resilient develop capacities to confront the hardships they face, implement novel solutions to problems, and handle difficult situations without emotional collapse. This doesn't mean they will always succeed, because in some cases they won't, but the fact that some people are flexible and persistent increases their success rate to the point that they can make significant advances in whatever they set their minds to. This aspect of human nature gives us reason to hope about the future. [13]

People who develop abilities to adapt to their environments can actually change them in order to meet their needs. This isn't without consequences, however, since people often concentrate on short term interests or goals, which impact major systems in the long run, often without considering later results. Anthropogenic changes—i.e. changes from human activities—have begun to impact our world on a global scale, modifying our atmosphere, hydrosphere, climates, and biosphere. We're still ignorant of what the long term effects might be, but measurements of changes in atmospheric contents, water acidity, oxygen content in the hydrosphere, and mean global temperatures are indisputable, as well as somewhat alarming. In spite of this, some people still argue whether or not humans are actually responsible for the impacts of those changes. [14] What they cannot deny are the consequences of those changes.

Regardless of who or what is responsible for changing global systems, people continue to adapt to a wide variety of conditions and circumstances, and are likely to do so in the future, barring cataclysmic events that would wipe out life as we know it.

77.

There's a universal law that can be compared to the stretching of a bow. What's longer will be shortened. What's shorter will be lengthened. Wise ones who use

Systems Thinking place the talents of their minds into service for humanity.

9.77—A question of balance

The world is filled with self-balancing systems. Although we may attempt to change or disrupt them, these systems have a tendency to counterbalance or adjust to anything that is done to them, and revert to a new balance.

Earth is a resilient self-balancing system. In spite of the fact we have exploited resources and produced significant impacts on this world's atmosphere, oceans and ecosystems, global dynamics have adjusted to those impacts, and in most cases have reestablished equilibrium.

There are limits to modifications we can make without serious consequences for all life forms, including ourselves. Ultimately, whatever we do will have lasting effects on our future existence, whether we recognize this or not. [15]

This form of global karma is inevitable. Is there really another world we could move forward or back to? Either we learn how to manage our lives with the resources of our planet, or we might as well forget about going elsewhere, or going anywhere at all. Our challenge is to find harmony and equilibrium in the systems we interact with.

Otherwise, we face the consequences of our own behaviors if we upset the global dynamics of our planet.

<h2 style="text-align:center">78.</h2>

Nothing in the world flows more freely than water, and there's nothing like water to erode what is powerful and strong. Though this is something people may know, it's a principle that many fail to practice.

9.78—On the nature of water and human behavior

The nature of water is surprising and fickle. There's nothing like a placid lake or pond to calm the soul, or the overwhelming power of a *tsunami* to terrify us. Water has qualities and exhibits contradictory behaviors, not unlike our own. In terms of contradictory actions: people are capable of immense love and altruistic behaviors; they are also capable of unspeakable hatred, violence, and cruelty, along with other horrific human behaviors. Is it any wonder we become confused sometimes when we try to deal with our own social problems?

We can observe changing states of water everywhere—solid as ice, liquid in living organisms, lakes, rivers, and oceans, and gaseous in our atmosphere or combustive conditions. Water is one of the most versatile and changing compounds we could ever encounter, without which we wouldn't exist. Hydrogen and oxygen, as elements, are more reactive alone than in water as a compound.

Changing states in the nature and behavior of human beings among all species of this planet are equally ing. Homo Sapiens is the most versatile and changing creature of this world. This hasimpacted our planet (and one another) in numerous positive and negative ways.

Changing states of human behaviors among people are sometimes perplexing, given that we demonstrate love and solidarity in our families and communities, but may betray one another, or fight with our neighbors for reasons which are not justifiable, other than because of fear, greed, or selfishness. We behave differently as individuals or as groups in societies. And although we justify certain behaviors and subsequent consequences, these aren't easily reconciled since there's a duality of individual selfishness that contrasts with group cooperation, as these behaviors evolve over time. [16]

A difference between this planet we inhabit and ourselves, however, is that while one is absolutely essential to the existence and well-being of life, it isn't so clear we are essential to anything else except ourselves, and even then, there's some doubt from the behaviors we exhibit at times with one another.

Nothing is quite what it seems. Many system behaviors are counterintuitive, including human behaviors. The ways systems actually work often turn out to be the opposite of what we expect. Housing programs produce higher concentrations of populations with more unemployment (but so would slums). Drug control operations to reduce the availability of drugs often produce higher crime rates. Social programs of popular governments deplete public finances through deficits which create economic recession. Whatever we do has unanticipated consequences. New developments may also arise from creative destruction. [17]

Before we leave this on a negative note, it's worthwhile to notice that people have begun to recycle wastes and reutilize what once was discarded. We have stepped back from nuclear destruction, we have greater respect for social and cultural diversity, and we've also begun using renewable forms of geothermal, hydroelectric, solar, and wind energy. Although we have responded with these improvements for a relatively short period of time, and dedicated only a small fraction of our efforts to these issues, we start from wherever we are standing. [18]

Those who are aware will be more prepared to face the unexpected. People who admit their mistakes and learn from them have already started to resolve problems we may encounter, now and in the future. Experimenting goes beyond the laboratory, to a much greater venue that encompasses humanity as a whole. We're discovering what works for the future of our species.

79.

Curing a wound leaves a scar. Though a dynamicist may expose the worst part of systems analysis, he can do this without arguing or placing extra burdens on everyone else.

9.79—The paradoxical path to improvement

A path to improvement starts by admitting we have something to improve. So it follows that improvement implies change—for without that we cannot initiate anything. And to change we need direction, energy, and commitment.

What's implied by something we can change or improve? It's often discovered from finding the root causes for problems or mistakes—often those we are making ourselves. But it's tough to be decisive when we first must admit we're wrong to ourselves, because without that, we cannot make the corrections we need for improvement. [19]

This is where ego comes into play, since people don't enjoy admitting mistakes or their roles in making them (but they do blame other people since it's easier to see errors which others make). Why should we admit our own errors?—Because otherwise, we're condemned to repeat them.

Repeating the same error is worse than making a mistake once and correcting it. At least the second option gives us an opportunity for improvement. This demonstrates that following a path for improvement won't be possible unless we're reflective about our own behaviors and willing to change, admit our own errors, and take necessary steps to resolve them. When you think about it, what is there to be proud about when you make mistakes?

So it's better to be humble and correct errors than to be proud of making them. And yet many people do exactly that. What's strange about counterintuitive systems is that we often get better results when we do the opposite of what others want or expect us to do. [20]

For people who are impressed by results, however, problems are important. What if we could eliminate potential problems before they begin? Our "results oriented" society would have no evidence to present for improvement. So the modern world thrives on problems and problem solvers. The two go hand in hand, just as "good" and "evil" define one another by comparison.

Perhaps you can see where this is going. Our societies tend to generate as many problems as they solve, so we have the endless march and material for improvements. We're constantly occupied with producing more results without actually ever achieving an end state of perfection. Thank Gaia that the overall energy system of our planet is renewable.

We wouldn't know what to do if we actually solved our problems once and for all, because the problem solvers and other related professions (and there are many) would be left with nothing to do, get paid, or be recognized for. Strange as it may seem, our world is set up to treat the symptoms of problems without ever solving them. And the people who solve temporary issues wouldn't appreciate being left unemployed. If we were able to stop crime, should we fire the police force?

So it goes without saying that people could find other things to do, but try telling them they could change professions and see what happens. As mentioned before, real problem solvers wouldn't be appreciated because of the paradox that no one is ever recognized for solving problems which never happened.[21] Where would a savior be if there was no one to save?

Unfortunately, we will feed on our problems until we learn somehow to let go of our obsessions. Thus we make weapons as a deterrent to others who make weapons to deter us. We conquer territories that have no need to be conquered. A few people take more to become rich and leave others without what they need. Circular reasoning has the advantage of being understandable to those who only see one side of an issue. [22]

80.

A small country is subjected to many systemic distortions. Actual boundaries of such systems are difficult to define, and are often inadequate. Small systems are often afflicted by the external influences of larger ones.

9.80—*Minor mentalities*

What can we say about minor mentalities? Although people could focus on greater things, village minded, egocentric citizens tend to lose touch with the roots of who they really are and focus on inconsequential activities, as if these were important. They lose sight of other things which are critical on a larger scale and could uproot their lives if left unattended or ignored (think of California Burning and the Flood of New Orleans). [23] [24]

Although a village mentality seems nice, it isn't so nice when we ignore that Earth is a global community which all of us share. This goes beyond personal concepts of beliefs and charity. Other people are reflections of ourselves.

—We're not practicing kindness by allowing poverty.

—We're not fighting for liberty and justice if we take up arms against one another.

—We're not practicing real democracy without equitable economic opportunities.

—We have no moral authority if we wage wars which kill other human beings, allow people to die of starvation, or refuse to care for the old, infirmed, or sick.

Although people may think specifically in terms of beliefs, neighborhoods, or personal interests, we need to think in larger terms for a respectful coexistence on our planet as human beings. This counts for more than us as individuals or our species alone. We also need to develop a healthy respect for all living specie because we are as much a part of life, as it is a part of us. [25]

Peope somehow ignore what we need most—*we need to include everyone.*

Thinking Globally, Acting Locally

—Earth already has the resources everyone needs. We only sell as a means to redistribute them.

—Distribution isn't rocket science, but we call it economics. What others might call most versions of economics: *enlightened selfishness.*

—Once we invented money as a means of exchange, we never looked back.

—Banks are the *New Religion of the Ages.* People flock to their temples every day, except Sundays, out of respect for other religions.

—Branding is the ultimate way of saying, "I've got this but you don't." So what does that imply about equitable distribution?

—Education taught us to compete when we could have learned to do otherwise.

The biggest questions for many:

What do others have that's worth competing for?

How can I get more?

—Governments shouldn't be divisive, but since we discovered politics, democracy has popularized empty rhetoric, partisanship, and unending political arguments.

—If governments actually worked as they were supposed to, without self-serving civil servants, where would we be now?

—We should be able to see from the preceding comments that the saying "to give is better than to receive" is not just charity—it's common sense.

81.

Sincere words are not agreeable. Agreeable words are not sincere. People who ask for help may blame the messenger, but the analysis of a system isn't meant to be negative. A wise person knows how to explain even what is difficult, to accept and at the same time avoid conflicts.

9.81—Fully aware

How many times have we heard speeches filled with agreeable phrases and learned to ignore them? Or when we have heard the vociferous denunciations of politicians, we've ignored them as well? Would we do this if they were seriously intended? The rhetoric of politics is not to be taken seriously. And yet, it's sad that we then ignore what requires our greatest attention. We ignore that we need to become conscious and aware as a species, not just as a few isolated individuals—not as a single Mahatma Gandhi—but ALL of us.

It's not about 'the smartest guys in the room.' It's about what we can do collectively. So the intelligence that matters is collective intelligence, and that's the concept of 'smart' that I think will really tell the tale.— Peter Senge [26]

The problem now is that most of us give lip service to what we have learned through knowledge, but rarely from what we have lived or learned through wisdom and experience. We may know something is true as individuals but feel powerless for social change. There is power, however, in all of us. And at some point, a critical mass of human awareness could produce the trigger for change.

Much like the chain reaction of a nuclear explosion, the time expended for certain changes might be almost instantaneous, and if that revolutionary moment ever arrives, we will only wonder then why it didn't happen sooner.

男女

Tao Te Ching

道德经

By Lao Tsu

Translated by Michael McCurley

This translation of *Tao Te Ching* is taken from a Spanish version by Alfonso Colodrón

1.

The Tao can be spoken of, but not the eternal Tao.

It can be given names, but not its eternal name.

As the origin of heaven and earth, it has no name, but it is the Mother of all things that can be named.

Therefore, from what is always hidden, we can contemplate its internal essence.

But from what is constantly manifested, we contemplate its external presence.

Both from the same source, have different names, and are called mysteries.

The Mystery of mysteries is the door to all essence.

2.

When the world discovers what is beautiful as beauty, it becomes ugliness.

When the world discovers what is good as good, it becomes bad.

Certainly what is hidden and manifested, are generated from one another.

Difficulty and facility are complements of each other.

Long and short are manifested by contradiction.

High and low are established by mutual measurement.

Voice and sound harmonize together.

Forward and backward mutually succeed one another.

For this reason, a wise one handles his affairs without acting, and extends his teachings without speaking.

He denies nothing among innumerable things.

He builds without attributing, and works without accumulating anything.

He finishes his work without pride and because he is unselfish, no one can take it from him.

3.

When we do not exalt talent, people abandon rivalry and discord.

When we do not value what is rare, people do not take or steal.

When we do not exhibit what others covet, the hearts of people are serene.

For this reason, a wise one begins to govern by:

— emptying hearts of desires,

— nourishing hungers,

— debilitating ambitions,

— and fortifying strengths.

In this way, people remain without complaints or desires, and those who are wise are careful not to act.

Through practicing Inaction, everything will harmonize.

4.

The Tao is like an empty cup, which being used, can never be filled.

Being bottomless, it becomes the source of all things.

It dulls sharp blades, unties knots, harmonizes lights, and unites the world as one.

Hidden in the depths, it becomes eternal existence.

It ignores where it was born, but is the common ancestor and parent of all things.

5.

Heaven and earth are beyond feelings and transcend all that exists.

A sage also transcends, and is above the feelings of people as well.

Heaven and Earth are unified in the emptiness of space, which is inexhaustible. The more it extends, the more we find.

It is best to look within.

6.

The essence of the Source never dies.

We call it the Feminine Mystery.

The door to the Feminine Mystery is called the Root of Heaven and Earth.

Extending like strands of a spider's web, it only shows the traces of its existence;

But those who drink from this source will find it inexhaustible.

7.

Heaven is eternal and Earth abides.

What is the secret of its eternal living presence?

What lives always does not live for itself.

For this reason, a wise one who goes last is before everyone else.

By letting go of himself, he is sure to be safe and sound.

Disinterest is the realization of being.

Is it not by disinterest that being is realized?

8.

Supreme kindness is like water.

Water benefits all without fighting anything.

It goes to places people avoid.

For that reason, it is close to the Tao.

Having chosen its place, it never leaves the earth.

So in cultivating your mind, you need to dive deep.

In dealing with others, you need to be generous and kind.

In speaking, you need to measure your words.

In governing, you should know how to maintain order.

In administering, you should be efficient.

When you act, be sure you choose the right moment.

If you fight against no one, you are free from reproach.

9.

To maintain plenty, if only you could stop time!

If you continually sharpen and use a sword, its sharpness won't last.

If you fill your house with jade and gold, you will not always be able to protect it.

If you accumulate riches and honors, you will only reap calamities.

This is the law of Heaven and Earth:

Withdraw once you have completed your labor!

10.

To maintain your spirit and breath of life united, can you conserve their perfect harmony?

When you unify your vital energy to attain flexibility, have you achieved the state of a newborn infant?

When you purify and illuminate your inner vision, have you been cleansed of all impurity?

When you care for people and govern your state, are you capable of governing with intelligence?

In opening and closing the doors of heaven, can you employ your feminine side?

Illuminated and in possession of an ample and penetrating vision, can you also remain inactive and detached?

Breed your people!

Feed your people!

Raise your people without claiming credit!

Do your work without accumulating anything!

Be a leader, not a butcher!

This is called hidden Virtue.

11.

Thirty spokes converge upon a single hub;

From the center axle, depends the use of a chariot.

Although we make a pot from a mass of clay, what is useful is its inner space.

We build doors and windows of a room, but these empty spaces make it habitable.

In this way, what is tangible has its advantages, but we find utility from what is intangible.

12.

Five colors blind the eye.

Five notes deafen the ear.

Five flavors dull the taste buds.

Speed and the hunt addle the mind.

Precious objects tempt man to do evil.

Because of this, a wise one cares for the stomach, not the eye,

He prefers what is within, instead of outside him.

13.

Take failure as an agreeable surprise, and calamity as you would your own body.

Why should we accept failure as an agreeable surprise?

Because a humble state is a favor: falling into it can be agreeable, and so can starting over again.

Why should we esteem calamity as we would our own bodies?

Because our bodies provide the same source for these calamities.

Without bodies, what problems could there be?

So then, only one who is willing to give up his body to save the world, deserves the confidence of the world.

Only one who can do this with love deserves to be administrator of the world.

14.

Look, but you cannot see.

Its name is Formless.

Listen, but you cannot hear.

Its name is Inaudible.

Try to catch it, but it is untrappable.

Its name is Incorporate.

These three attributes are impenetrable;

And because of this, the three are one.

Its upper part is not luminous,

Nor its inner portion dark.

What continuously flows from the un-nameable,

returns to what is beyond the reign of existence.

We call this Form without form, Image without images.

We call it the indefinable, the unimaginable.

Turn to face it, and you will not see it!

Look, but you cannot follow.

So, given the immemorial Tao,

you can manage the realities of the present.

Knowing origins is to be initiated in the Tao.

15.

The ancient disciples of the Tao were subtle, flexible, profound, and global.

Their minds were too profound to be penetrated.

Being impenetrable, we can only vaguely describe what they appeared to be.

They were bold like one who crosses a river current in winter, shy like those who fear their neighbors around them;

Prudent and courteous as guests, transitory as ice, which is about to melt, simple like an unsculpted trunk, deep like a cavern, confusing like a swamp.

And yet, who among other people, can pass from what is cloudy and uncertain to clarity?

And who, if not these, can pass slowly but surely to bring what is inanimate to life?

One who observes the Tao does not desire to be filled.

So, more precisely, if you are never full, you can be maintained always as a hidden seed that will not sprout and suddenly overgrow.

16.

Reach the supreme Emptiness.

Embrace inner peace with a decided heart.

When everything becomes agitated at once, just contemplate the Return.

To blossom as everything does, you need only return to your roots.

Returning to your roots is the way to find peace.

Finding peace is to realize your own destiny.

Realizing your destiny is to become eternal.

And knowing the Eternal is called Vision.

If you do not know the Eternal, you will fall blindly into disgrace.

If you know the Eternal, you can understand and do anything.

If you can understand and do anything, you are capable of doing justice.

To be just is to be like a king, and to be like a king is to be like the sky.

To be like the sky is to be one with the Tao; and being one with the Tao is to last forever.

One like this will be safe and sound, even after his body has disintegrated.

17.

The best governor is one who people barely notice.

After this comes the one who people love and adore.

And then comes the one people fear.

Last is the one who is despised and defied.

If you are untrusting, others will not trust you.

The wise one passes by unnoticed and guards his words.

When his work is completed and all things are done, everyone will say: We are the ones who have done this!

18.

When the Great Tao was abandoned, then there appeared benevolence and justice.

When intelligence and craftiness emerged, then hypocrites appeared.

When six family relations lost their harmony, then brotherly piety and paternal love appeared.

When darkness and disorder began to reign in the country, then loyal functionaries appeared.

19.

Renounce knowledge, abandon ingenuity, and people will profit immensely.

Renounce the benevolence and justice of civilization, and people will return to their natural sentiments.

Renounce craftiness and smartness, and thieves and evildoers will disappear.

These three furrows of the Tao are not sufficient unto themselves.

Because of this they are subordinated to a greater principle:

See what is Simple and embrace the Primordial!

Diminish the ego and moderate desires!

20.

When we abandon what we have learned, contradictions will disappear.

What difference is there between "this" and "that"?

What distinction can be made between "good" and "bad"?

Why fear what others fear?

What senselessness!

When all the world is happy and smiling, as if they are celebrating the sacrifice of a bull, and as if they have gone to the Spring Pavilion, only I remain passive and tranquil like a newborn which has never smiled.

Only I am homeless, like one who has no home to return to.

While all the world lives in abundance, only I seem to possess nothing.

How crazy I am!

What a confused mind I have!

Everyone is brilliant, so brilliant!

Only I am dark, so dark!

Everyone is smart, so smart!

Only I am quiet, so quiet!

Liquid like the ocean, without purpose like the gusts of a storm.

Everyone is channeled in their cause, while I am sidelined and obstinate.

But the one way I am really different from everyone else is knowing I have sustenance from my Mother.

21.

The nature of Grand Virtue is found by following the Tao and only the Tao.

But what is the Tao?

It is something divergent and imprecise.

Divergent and imprecise!—but it contains Form in its interior.

Divergent and imprecise!—but it contains Substance in its interior.

Grave and obscure!—but it contains the Seed of Vitality in its interior.

The Seed of Vitality is quite real: it contains inexhaustible Sincerity within.

Throughout the ages its Name has been preserved to remember the Origin of all things.

How will I know the nature of this Origin of all things?

Because it is within me.

22.

Bow and you will be made complete; bend and you will be straightened.

Maintain emptiness, and you will be filled.

Grow older, and you will be renewed.

If you have little, you will gain much.

If you have much, you will become confused.

Because of this, a wise one embraces Unity and becomes a Model for everything under heaven.

When you do not show off, you will shine; if you do not justify, you will be known for that; if you do not proclaim your abilities, people will trust that; if you don't exhibit your achievements, for that reason they will remain.

Don't compete with anyone, and nobody will compete with you.

Certainly, the words of this old refrain are true: bow and you will be complete.

What is more, if you truly achieve this plentitude, everything will come to you.

23.

Only simple and tranquil words will come of age by themselves.

A tornado cannot last an entire morning, nor a downpour an entire day.

From where have these come?—from Heaven and Earth.

Not even Heaven and Earth make these phenomena last very long; what should we expect from the precipitated efforts of human beings?

For this reason, one who cultivates the Tao is one with the Tao; one who practices Virtue is one with Virtue; and one who courts Loss is one with Loss.

Becoming one with the Tao is a welcome access to the Tao.

Becoming one with Virtue is a welcome access to that Virtue.

Becoming one with Loss is a welcome access to that Loss.

A lack of confidence on your part, leads to a lack of confidence in everyone else.

24.

One who stands on tip-toes cannot sustain this.

One who sustains the hanged, cannot walk.

One who exhibits himself cannot shine.

One who justifies himself cannot obtain honors.

One who exalts his own capabilities has no merit.

One who praises his own achievements will not last.

In the Tao, these things are called spoiled food and malignant tumors, which are abominations for all beings.

For this reason, a man of the Tao does not put these things into his heart.

25.

There is Something undefined, but complete in itself, born before Heaven and Earth.

Silent and unlimited, unique and immutable, although it impregnates everything without exception—we could consider this the Mother of the world.

Since I don't know her name, I call this the Tao, and for lack of better words I call this: The Great.

To be great is to move forward, to move forward is to go far, and going far is also to return.

For this reason, the Tao is Great, as Heaven and Earth are Great, as the King of creation is Great.

So man is guided by the laws of the Earth.

The Earth is guided by the laws of Heaven, the Heavens are guided by the laws of the Tao, but the Tao is guided by its own laws.

26.

Heaviness is the root of lightness.

Serenity is the owner of unquietness.

For this reason the wise one who travels all day, never leaves his caravan, though there may be wonders to be seen, he remains tranquil in his own house.

Why would a man of ten thousand cars expose his lightness to the world?

To behave lightly is to separate yourself from your own roots; becoming agitated is to lose dominion of yourself.

27.

Walking on the correct path leaves no trail behind you; correct speech leaves nothing to be criticized, good calculation does not require a calculator; a good lock does not require a latch or slide-bolt—but nobody can open it; what is well tied does not require cords or knots, but no one can untie it.

For this reason, a wise one always knows how to save people, so nobody is abandoned; he always knows how to save things, so nothing is wasted.

This is called: following the guiding Inner Light.

For this reason, better people are teachers of ones who are less so, and these will take charge of them.

To not be happy with your own task is to follow an incorrect path, no matter how intelligent it may seem.

This is an essential principle of the Tao.

28.

Know the masculine, stay in the feminine, and *be* the Way for the World.

To be the Way for the World is to walk constantly the path of Virtue, without straying from it, and returning anew to infancy.

Know the white, stay in the black, and *be* a Model for the World.

To be a Model for the World is to walk constantly the path of Virtue, without missing a single step, and returning anew to the Infinite.

When you know the glory, keep humble and *be* the Source for the World.

To be the Source for the World is to live a fertile life of Virtue, and to return to Primordial Simplicity.

When Primordial Simplicity is divided, it becomes useful containers, which in the hands of a Wise One, become functionaries.

For this reason, a great tailor makes few cuts.

29.

Could someone really conquer the world and do what he wants with it?

I don't see how that person could succeed.

The world is a sacred vessel which cannot be manipulated or dominated.

To manipulate it is to ruin it, and to dominate it is to lose it.

For this reason, there is a time for things to move forward, and a time for them to move back; a time to breathe slowly, and a time to breathe quickly; a time to grow in strength, and a time to decline; a time to rise, and a time to fall.

Because of this, the Wise One avoids extremes, excesses, and extravagances.

30.

One who knows how to govern does not attempt to do so by force of arms.

The nature of military weapons tends to revert against those who use them.

Where armies encamp, grow thorns and brambles.

A great war will invariably produce bad years to follow.

Those who wish to protect their own state effectively should not attempt to expand it.

When you achieve your purpose, you should not exhibit triumph, or boast about your capacity, nor feel proud; instead you should lament that you couldn't prevent war in the first place.

You should never think about conquering others by force.

This is because expanding excessively can precipitate collapse, and this is against the Tao, and what is against the Tao will soon cease to exist.

31.

Sophisticated weapons of war precede calamity.

All things and beings hate them.

For this reason, a person of the Tao does not put these things in his heart.

In daily life, a person of nobility considers his left as a place of honor.

In war, the right is the place of honor.

Since weapons are instruments of misfortune, these are not adequate instruments for nobility.

These are only used out of absolute necessity, because peace and calm should be what your heart desires most, and for that, every victory is no reason to rejoice.

To be happy about such victory is to be happy about the killing of human beings!

For this reason, a man who is happy about killing other people, cannot hope to prosper in the world of human beings.

In festive occasions the left is preferred, but for unlucky ones it is the right.

This means that war can be compared to a funeral service.

When many people have been killed, it is only just that the survivors cry for the dead.

For this reason, every victory is also a funeral.

32.

The Tao eternally lacks a name.

Although small in its Primordial Simplicity, it is not inferior to anything else in the world.

If only a governor could have this, everything would give homage.

Heaven and Earth would harmonize, and a soft dew would fall.

Peace and order would reign among all peoples, without any need for superior orders.

Once Primordial Simplicity was divided, different names appeared.

Aren't there already enough names today?

Hasn't the moment come to end this now?

Knowing when to do this will preserve us from danger.

The Tao is to the world like a great river causeway, or what the ocean is to rivers and streams.

33.

One who knows others is intelligent.

One who knows himself has inner vision.

One who conquers others has strength, but one who conquers himself is really powerful.

One who knows he has obtained much is rich, but one who assiduously follows the Tao is someone with constant purpose.

One who stays in the place where he has found his true home will live long, and one who dies but does not perish, will enjoy authentic longevity.

34.

The Great Tao is a universal current.

How could it be diverted to the left or to the right?

All creatures depend upon it, and it is never denied to anyone.

It carries out its labor, but does not attribute itself.

It feeds and clothes everyone, but never Lords this over anyone.

In this way, it can be called the Small.

All things return to it, as to their homes, but without owning them.

In this way it can be called The Great.

And because it has no desire to be Great, its greatness is fully realized.

35.

One who has the Great Symbol attracts all things to it.

What they fully trust will never do harm, and in it they will find peace, security, and happiness.

Music and delicacies may make a guest stop for a while.

But the words of the Tao have lasting effects, even when these are bland and without flavor, even when they do not attract by sight or sound.

36.

What in the end must contract, must first expand.

What in the end must weaken, must first be strengthened.

What in the end must be discarded, must first be prized.

What in the end must be dispossessed, must first be enriched.

Here lies a subtle wisdom about life:

The soft and weak will triumph over what is hard and strong.

The same for a fish which must not abandon the depths, a governor must not show his resources.

37.

The Tao never carries out a single action, yet leaves nothing undone.

If a governor could follow this, all things would develop by themselves.

When what is developed begins to become agitated, it is time to keep things on track with the help of unnamable Primordial Simplicity; only this can moderate human desires.

When the desires of people are moderated, we will find peace, and the world will harmonize on its own.

38.

Maximum Virtue is not virtuous, so it has virtue.

Minimum Virtue, never free of itself to be applied, has no virtue.

Maximum Virtue is not ostentatious, nor does it have personal interests to serve.

Maximum compassion *is* ostentatious, but it serves no personal interests.

Maximum morality is not only ostentatious, but also serves personal interests.

Maximum ritual which becomes ostentatious, offers no answers, so it must be imposed by force.

When one loses the Tao, one falls back to virtue.

When one loses virtue, one falls back to compassion.

When one loses compassion, one resorts to morality.

When one loses morality, one resorts to ritual.

So then, ritual is only the appearance of faith and loyalty; it is the beginning of all confusion and disorder.

Prescience may be the flower of the Tao, but can be also the beginning of stupidity.

For this reason, one who is completely realized puts his heart in substance more than appearance, and in its fruit more than the flower.

He sincerely desires more from what is within than without.

39.

All ancient things became what they were by achieving Unity.

The heavens achieved Unity and became diaphanous; the earth achieved Unity and became tranquil; spirits achieved Unity and were filled with mystical powers; the springs achieved Unity and were filled; ten thousand creatures achieved Unity and could reproduce; Lords and princes achieved Unity and became reigning governors of the world.

All of these are what they are by virtue of this Unity.

If the heavens were not diaphanous, they would explode into pieces; if the earth were not tranquil, it would collapse into ruins; if the springs were not filled, they would run dry; if spirits were not filled with mystical powers they would cease to exist; if lords and princes were not sovereign, they would stumble and fall.

In truth, the humbleness of the root blossoms into greatness, and what is lofty is built upon humble foundations.

For this reason, lords and princes call themselves the Invalid, the Ignorant, and the Indigent.

Perhaps they will also discover that they depend upon what is humble.

In truth, great honor is equivalent to no honor at all.

It is not for sages to shine like jade, better like the roughest of stones.

40.

All movement of the Tao consists in return.

Softness is the procession of the Tao.

All things under the heavens have been born from what is corporeal.

And what is corporeal has been born from the incorporeal.

41.

When a sage hears what is said of the Tao, he practices with diligence.

But when a mediocre scholar hears what is said of the Tao, he oscillates between faith and incredulity.

When a scholar without values hears what is said of the Tao, he laughs hysterically about it.

But if people didn't laugh at it, the Tao wouldn't be the Tao.

The sages of antiquity have truly said:

The Way which is clear, seems dark.

The Way which is progressive, seems regressive.

The Way which is soft, seems rough.

Superior Virtue seems to be a pit.

Great whiteness appears to be stained.

Exuberant Virtue appears incomplete.

Established Virtue appears tattered.

Solid Virtue appears to have melted.

The great Cube has no corners.

And great talents mature late.

The great Sound is silence.

The great Figure has no form.

The Tao is hidden and has no name, but only it knows how to help and complete.

42.

The Tao gave birth to One, and One gave birth to Two, and Two gave birth to Three, and Three gave birth to innumerable things.

These innumerable things are backed by the Yin and embrace the Yang, and their vital harmony is derived from an adequate mix of these two vital winds.

What could be more detested by men than invalids, ignorants, and indigents?

And yet these are what lords and princes have named themselves.

In truth, we can win by what we lose, and we can lose by what we have won.

Let me repeat what someone once taught me: a violent person will come to a violent end.

Whoever affirms this could be my parent and teacher.

43.

The softest of all things is superior to the most rigid of them.

Only nothingness penetrates where there is no space.

For this reason, know the advantages of Doing Nothing.

Few things are more instructive than the lessons of Silence, or as beneficial as the fruits of Doing Nothing.

44.

Which do you value more—your name or your body?

Which is more precious—your body or your health?

Which is more painful—gain or loss?

Know then that excessive love for anything will cost more in the end.

Accumulating too much property will cause you significant loss.

Knowing what is enough could make you immune to this disgrace.

Knowing when to stop could save you from dangers.

Only this way could you live for a long time.

45.

The greatest perfection may seem imperfect, but its exercise is inexhaustible.

The greatest plentitude may seem empty, but its function is everlasting.

The greatest rectitude may seem twisted.

The greatest ability may seem awkward.

The greatest eloquence may seem to stutter.

Movement may overcome the cold, but calm overcomes heat.

What is pacific and serene is the Norm of the World.

46.

When the world is in possession of the Tao, charger horses are taken out to fertilize fields with their excrement.

When the world is without the Tao, these beasts of war feed in the suburbs.

There is no greater calamity than not knowing when enough is enough.

There is no greater defect than covetousness.

Only when one knows there is enough will there always be plenty.

47.

Without leaving your door, you can know the affairs of the world.

Without spying out your window, you can know the Way to Heaven.

But the further you go, the less you will know.

In this way, the wise one knows without traveling, observes without looking, and achieves without acting.

48.

Learning consists in accumulating the Tao, day by day; practicing the Tao consists in reducing it day by day.

Continue reducing and reducing until you reach the state of Nothingness.

Nothingness, and yet there isn't anything left to do.

To win the world, you must renounce everything.

If you still have personal interests to serve, you will never win the world.

49.

The Sage has no personal interests, but makes the interests of others his own. He is kind with those who are kind, but also kind with those who aren't.

Because Virtue is kindness; he also trusts those who don't deserve trust.

Because Virtue is trust, in the midst of the world, the Sage is modest and shy.

To benefit the world, he maintains an open heart.

While all the world strains its eyes and ears, the Sage only smiles like an amused child.

50.

What is outside of Life, finds itself in Death.

Thirteen are the companions of Life; thirteen are the companions of Death. And when a living person penetrates the Kingdom of Death, he will also have thirteen companions.

Why is this?—Because he is so attached to the resources of Life.

It is said that one who knows how to live will not find tigers or wild buffalos in his path, and he will leave the battlefield without being injured by weapons of war. This is because there is no place for him to be affected by buffalo horns, nothing for a tiger's claws to tear, and no target for any weapon of war.

Why is this?—Because there is no place in him for Death.

51.

The Tao gives life, and Virtue sustains it.

Matter gives form, and the Environment perfects it.

For this reason, all things venerate the Tao and give homage to Virtue.

No one has ordered that the Tao be venerated, or that Virtue be given Homage. These things happen spontaneously.

It is the Tao that gives life.

It is Virtue that sustains it, causes it to grow, conserves it, protects it, supports it, feeds it, and covers it with its wings.

Give life without attributing anything, do your work without accumulating anything for it—be a leader, not a butcher—this is called hidden Virtue.

52.

Everything found under Heaven has a common Origin.

This Origin is the Mother of the world.

After knowing the Mother, we begin to know her children.

After knowing her children, we should return and preserve the Mother.

Whoever does this will not be endangered, even if their body is annihilated,

> Block all passages!

> Close all doors!

> And at the end of your days, you will not be exhausted.

> Open these passages!

> Multiply your activities!

And at the end of your days, you will be defenseless.

To see what is small is to have inner vision.

Preserving what is weak is to be strong.

Use lights, but return to your inner vision.

Don't attract damages upon yourself.

This is the way to cultivate the Immutable.

53.

If I possess the smallest seed of wisdom, I will walk the Great Path, and my only fear will be to stray from it.

The Great Path is flat and straight, but people prefer more torturous routes.

The farmyard is clean and well made, but the fields are uncultivated and full of weeds. And the granaries are completely empty!

Those who wear fine clothes and carry sharpened swords, possess greater riches than they can enjoy! These are heralds of robbery!

But of the Tao, what can they know of it?

54.

What is well planted cannot be uprooted.

What is tightly bound cannot be loosened.

Your descendants will make offerings to your ancestors, from generation to generation.

Cultivate Virtue within yourself, and it will become a true part of you.

If you cultivate this in a family, it will endure.

If you cultivate this in a community, it will live and grow.

If you cultivate this in a state, it will flourish abundantly.

If you cultivate this in the world, it will become universal.

For this reason, a person should be judged as a person, a family as a family, a community as a community, and a state as a state.

How can I know about the world?

By what is found within me.

55.

One who is anchored in Virtue is like a newborn infant.

Wasps and venomous serpents will not sting, nor ferocious animals attack, nor will rapacious birds hurl themselves upon him.

His bones are soft, with bland tendons, and yet he will be firmly attached.

He has not known the union between male and female, but grows in full plentitude, and conserves his vitality with perfect integrity.

He shouts and cries all day without becoming hoarse, because he incarnates perfect harmony.

Knowing that harmony is to know the Immutable.

Knowing the Immutable is to have inner vision.

But to precipitate the growth of life is nefarious.

Voluntarily controlling respiration is to violate it.

Overgrowth is to become old.

All of these are contrary to the Tao, and what is contrary to the Tao will soon cease to exist.

56.

One who knows does not speak.

One who speaks does not know.

Block all passages!

Close all doors!

Chip all blades!

Untie all knots!

Harmonize all lights!

Unify the world as one!

This is the Total Mystery, which cannot be courted nor rejected, benefited nor damaged, raised nor humiliated.

For that reason, it is the most Elevated in the world.

57.

A kingdom which is governed by ordinary laws, makes war by extraordinary movements, but wins the world by leaving it alone.

How do I know this?—Because of what I find within me!

The more Taboos and prohibitions there are in the world, the poorer people become.

The sharper the arms are that people possess, the more confusion will reign in a country.

The smarter and more astute people are, the more frequently strange things will occur.

The more complete promulgated laws are, the more evildoers and thieves will appear.

For this reason the Sage says:

I do not act, and people will transform themselves.

I love quietude, and people naturally find their own way.

I do not begin any new enterprise, and people will prosper.

I have no desires, and people return to Simplicity.

58.

When a governor is discrete, his people are simple and happy.

When a governor is cunning, his people are astute and unhappy.

Bad fortune is what good fortune brings, and good fortune is what bad fortune hides.

Who knows a definite end to this process?

Isn't there a norm for justice?

But when what is normal suddenly becomes abnormal, and what is favorable becomes a bad sign, people will face a dilemma.

For that reason, the Sage squares things without cutting, sculpts without disfiguring, straightens without force, and clarifies without darkening.

59.

To govern your people and serve heaven, there is nothing like sobriety.

To practice sobriety is to turn back before walking in the wrong direction.

To return before walking in the wrong direction is to possess a double reserve of Virtue.

To have a double reserve of Virtue is the best of all things.

To be better than all things is to reach an invisible height.

Only he who achieves this invisible height could have a kingdom.

And only he who has reached the Mother of the kingdom will last very long.

This is the way to become deeply rooted and firmly established in the Tao, the secret to long life and enduring vision.

60.

Governing a great kingdom is like cooking a small fish. When one governs the world according to the Tao, demons will lack spiritual powers. Not only will demons lack spiritual power, but other spirits will have no power to hurt people. Not only will these spirits have no power to harm people—even the sage himself will not harm his people. If only a government and its subjects would abstain from mutually doing harm to one another, then all the benefits of life would come to that kingdom.

61.

A great kingdom is like a valley in which all rivers converge. It is the Deposit for all that exists under heaven, the Feminine of the world.

The Feminine always conquers the Masculine by means of quietude, and lowering herself to him.

For this reason, if a great country lowers itself before a smaller one, it will win; if a small country lowers itself before a larger one, it will win. The first wins by inclining itself, the second by staying humble.

What the larger country wants is simply to encompass more people, and what the smaller one wants is to serve its protector. In this way, each of them gets what it wants, although it is up to the larger country to remain humble.

62.

The Tao is the hidden deposit of all things.

It is a treasure for an honorable person; it is a safeguard from error.

A good word has its own market.

A good work can be a gift to someone else.

But a man who has erred from the path is no reason for him to be set aside.

For that reason in the throne ceremony for emperors, or the naming of three ministers, let the rest offer their jade disks, preceded by cavalcades of horses.

It is better for you to offer the Tao without moving your feet.

Why do the ancients value the Tao?

Isn't this because, by its virtue, he who seeks will find, and guilt is forgotten?

For this reason, it is an unmatched treasure for the world.

63.

Practice Stillness.

Force yourself not to force anything.

Taste what has no flavor.

Favor the humble.

Multiply the little.

Reward injury with kindness.

Cut the problem before it appears

Plant what is great from a small seed.

Difficult things of this world can only be approached when they are easy.

The great things of the world can only be accomplished by paying attention to small beginnings.

So then, the Sage never has to fight with open arms against great things, although only he is capable of doing so.

One who promises too lightly is not trustworthy.

One who thinks everything is easy ends up finding everything is difficult.

For this reason, the Sage who considers things to be difficult from the beginning will not find them difficult in the end.

64.

What is at rest is easy to retain.

What is not manifested in omens is easy to predict.

What is fragile is broken easily.

What is small is easy to disperse.

Resolve problems before they appear.

Cultivate peace and order before confusion and disorder prevail.

A tree with a girth as wide as a man's arms is born from a tiny sprout.

A tower of six floors begins with a mound of earth.

A voyage of a thousand leagues starts from where you're standing.

One who acts in any affair can lose it.

One who holds onto anything loses it.

A Sage never acts, and so, never loses anything.

He doesn't hold onto anything, so he never loses it.

In managing your affairs, people may block them just as they are about to be finished.

Give your total attention from the start, and with patience until the end, nothing will be lost.

For this reason, the Sage desires to be free of desires, does not covet possessions that are difficult to attain, learns to unlearn what has been learned, and induces the masses to return to where they came from.

He helps all creatures to find their true nature, but doesn't lead them by the nose.

65.

In ancient times, those who were versed in the practice of the Tao did not attempt to instruct people, but instead kept them in a state of simplicity. Then why are people so difficult to govern? Because they are too intelligent! For that reason, one who governs his state by means of intelligence is an evildoer, but one who governs it without intelligence is its benefactor. Knowing these principles is a norm and a means. Maintaining a norm and a means constantly is what we call Mystic Virtue. Vast and profound is that Mystic Virtue. This takes all

things back to their return, until they reach the Great
Harmony!

66.

How could the sea be the King of all rivers?

—Because it lies beneath them!

For that reason it is the king of all rivers.

Because of this, the sage governs people by using
humble words; he directs by standing behind.

So then, when a sage is over people, they won't feel
his weight, and when he is in front, nobody feels hurt.

For this reason, everyone in the world is happy to
facilitate his progress without tiring of him.

Since he fights with no one, no one will ever fight
with him.

67.

Everyone says my Tao is great, although it appears
very strange to the world. But that is simply because my
Tao *is* great, since it appears like nothing on this earth! If
it were comparable to anything on earth, how small it
would appear from the beginning!

I have Three Treasures that I cultivate and guard
closely. The first is *Compassion*, the second is
Moderation, and third is *Make no attempt to be first in
the world*. Because—if I am compassionate, I can be

brave. If I am Moderate, I can be generous. If I make no attempt to be the first, I can be captain of all ships.

When a person wants to be brave without first being compassionate, generous without being moderate, or be a leader without first following others, he is only courting death!

Compassion alone can help you win a war. Compassion alone can help you guide a state. This is because Heaven comes to the rescue of those who are compassionate and protects them with *its* compassion.

68.

A good soldier is never aggressive. A good warrior is never irascible.

The best way to conquer your enemy is to beat him without confronting him.

The best manner to employ someone is to serve under his orders.

This is called the virtue of not-fighting!

This is called employing the capacities of men!

This is called being married with Heaven forever!

69.

Strategists have a saying: *I don't dare to be the host, instead I am the guest. I don't try to move a centimeter, but prefer to retreat a step.*

This is called advancing without moving, rolling up your sleeves without exposing the elbows, capturing an enemy without confronting him, and keeping an invisible weapon.

There is no worse disgrace than underestimating the power of your enemy.

To underestimate the force of your enemy is to lose your treasure.

For that reason, if you encounter troops in the field of battle, victory belongs to the side which is most afflicted.

70.

My words are very easy to understand, and easy to practice.

Although no one in the world practices or understands them, my words have a Predecessor; my words have an Owner.

People do not know.

For that reason they do not know me.

The fewer people who know me, the more noble are those who follow me.

For that reason, the Sage wears rough clothes, while he guards jade on his chest.

71.

Discovering our knowledge is ignorance is noble inner comprehension.

Considering our ignorance as knowledge is mental illness.

Only when we tire of our sickness, will we stop being ill.

Only when we stop being ill, will we tire of our sickness.

The Sage does not become ill, because he is tired of that sickness.

This is the secret to health.

72.

When people stop fearing you, it is a signal you are attaining great power.

Don't interfere lightly in their homes, nor impose heavy tributes upon them.

If you only stop knocking them down, they will be willing to fall for your cause.

For this reason, the Sage knows himself, but isn't overly proud; he loves but does not praise himself.

He prefers what is within to what is outside of him.

73.

One who is recklessly brave will perish; one who is brave without being reckless will survive.

There are two classes of valor. One is beneficial and the other is damaging.

Some things are detested by Heaven.

So who would know the reason?

Even a wise one can be disconcerted by such a question.

The Way to Heaven is to conquer without fighting, attract people without calling them, and to act according to plans without rushing them.

The Web of Heaven is crisscrossed with ample nets, and yet nothing escapes them.

74.

Once people no longer fear death, why would they fear its specter?

The more you make people fear death, the more they will violate the law. And you could with reason detain and execute them—then who would dare violate the law?

Isn't there always a Grand Executioner to kill?

Killing for the Grand Executioner is like cutting wood for a master carpenter, and of course, you would be fortunate if you don't injure your own hand!

75.

Why do people die of hunger?

—Because those above them tax them excessively.

For that reason they are dying.

Why are people so difficult to govern?

—Because those above them intervene too much, and serve their own interests.

For that reason people are difficult to govern.

Why do people take death so lightly?

—Because those above them live lives of luxury.

For that reason they take death lightly.

These people simply don't have anything to live for!

They know there are better things to do than to live life like that!

76.

When a person is alive (s)he is soft and flexible.

When dead, (s)he is hard and rigid.

When a plant is alive, it is soft and flexible.

When dead, it becomes withered and dry.

For that reason, hard and rigid are companions of death, and bland and flexible are companions of life.

So then, a powerful army tends to fall by its own weight, the same as dry wood is ready for the ax.

What is great and powerful will fall; the humble and weak will be honored.

77.

The Law of Heaven could be compared to stretching a bow. What is above will sink, and what is below will rise. If the chord of the bow is too long, it is shortened; if it is too short, it is lengthened.

The Law of heaven diminishes what is excessive and completes what is insufficient. The law of man is different: it takes from what is insufficient and gives to what is excessive. Who except a man of the Tao would put his excess riches in service to the world?

For that reason, the Sage accomplishes his work without accumulating anything and realizes his labor without becoming attached to it. He doesn't desire for his own merits to be seen.

78.

Nothing in the world is softer or more gentle than water; but there's nothing like water to erode what is hard and strong!— So nothing can replace it.

What is weak beats what is strong and what is soft beats what is hard. This is something everybody knows, but nobody practices.

For that reason, the Sage says:

Receiving the trash of a country is to be the lord of its temples.

Taking the disgraces of a country is to be a prince of the world.

Certainly, the truth seems to be its opposite!

79.

When a great wound is cured, it will always leave a scar.

How could that be desirable?

For that reason, a Wise One, even when he has the worst part of an agreement, always fulfills his convened part, and does not quarrel with others.

A virtuous person fulfills his duty; a person without virtue only places burdens upon others.

The Path to Heaven lacks personal effects, but it is always in harmony with generous people.

80.

How difficult it is for a small country with a limited population! Although it possesses efficient mechanical apparatus, the people don't use them. They stop worrying about death and abstain from emigrating to places farther away. There may be cars and ships, arms and shields, but no occasion to use or exhibit them.

It lets its people return to older ways of communicating. It procures to be content with its own food, gratified with its clothes, satisfied with its houses; it follows its own simple customs of life.

Although there might be another country in the vicinity, so close that both are within sight of one another, and both can reciprocally hear roosters crow or dogs bark, relations between them, throughout the lives of these two peoples, have nothing to do with one another.

81.

Sincere words are not agreeable. Agreeable words are not sincere.

Good people do not argue; those who argue are not good.

Wise people are not erudite; erudite people are not wise.

The wise one does not hoard anything for himself. The more he lives for others, the more complete his own life is.

The more he gives, the more he swims in abundance.

The Law of Heaven is to benefit, not to jeopardize.

The Law of the Sage is to fulfill his duty, not to fight against anyone.

Author's notes:

If you like this book, please share about it with someone else.

If you wish to change the world, start with yourself.

Systems Thinking could change our awareness as a species.

If you wish to learn more, dig deeper to see what you can discover.

The rest will naturally follow.

—

You can review *Tao Te Ching* for comparison and analysis. This is an excellent text for reflection and growth that has stood the test of time for centuries.

If you study further, you will discover there is still much to be explored. You will also find a virtual library of complete digital books and articles in the **Bibliography of Links and References** at the end of this book.

For more to read about Systems Thinking and other related topics, you may want to visit:

https://www.systemstao.com

I will be contributing additional articles and information.

Michael McCurley, San Jose, Costa Rica

Why Reconsider How We Think?

By Michael McCurley

We have a lot at stake, living in this world, learning to share and use its resources, besides working and completing with one another to survive.

At one time it may not have made much difference, but now that expanding human populations have covered nearly all of the inhabitable areas of this world, we have begun to impact our environment on a geological scale. The atmosphere, oceans, fresh water sources, and land masses have all been affected by our activities, as well as all other living species.

Although we may not be conscious how our day to day activities are affecting the future, they are, in fact, transforming our future possibilities with unimaginable consequences.

Why should this matter to us as individuals who are more focused on our own concerns, rather than more global issues? Why should any of this be important to us?

We're comfortable with what we know and believe because we think it won't change in the foreseeable future, which is fine as long as that lasts. But we are, in fact, changing our physical environment in ways that have produced accelerated degradation of our environment through exponential rates of production and consumption, which have added waste contaminants to our atmosphere, water supplies and land areas faster than these can be absorbed or assimilated over time. This also means we'll need more of limited resources that we're depleting, and unless we succeed in moving to renewable resources in time, we'll suffer from serious scarcities of food, water and energy resources that we need for the survival of our species. This is already happening.

Because we focus on our lives as individuals, we may fail to see the bigger picture—that we're part of a global society, which is still fragmented, competitive, and fearful of its own shadows. We may miss the fact that the impacts of our behaviors are beginning to change the dynamics of our climate, food production, economic development, and social interactions, which affect living standards for billions of inhabitants of our planet.

That consideration alone should motivate us to pause and reconsider where our behaviors are taking us. The world has become a reflection of how we act, which is based on our perceptions, how we think, and what we believe. And we know from history that what people have acted upon, thought, and believed has often been mistaken.

Although we focus on immediate time frames, we have created complex systems with extended scales, time frames, and delayed reactions which have impacts that extend far beyond individual lifetimes. These challenge our understandings and abilities to manage the complex results and side effects for what we've created. Understanding on a global level begins with individual awareness that we're the sum total of all our behaviors.

Perhaps you're wondering why people often do the opposite of what's needed for survival. You may also have noticed that governements aren't good at even following their own agendas and campaign promises. In spite of our human abilities to use reason and understanding, people are still generally quite emotional and irrational. This wouldn't be an issue except that the future depnds upon our behaviors, which are based on what we think, and yes, how we feel as well.

How we think plays a huge role in determining our future.

Beyond economics, politics, and religion, we need better understandings of how complex systems work and affect our lives—not just for the future, but right now, as well. We will discover, that system principles, like the Tao, are impartial and can be understood by anyone, even when they may appear at first to be contradictory. By understanding systems, we can deal more effectively with change and learn to recognize our own mistakes, which in turn will lead us to important discoveries and solutions for the issues that affect our lives. And as we advance in this process we will also come to appreciate our roles as members of a global society.

We will discover that how we think is the baseline of our existence because it determines our actions and behaviors. More importantly, our thoughts and perceptions establish who we are and what our destiny will be.

We think as we do because this works for us right now as individuals, but what if this doesn't work for us as a species? Most of us recognize we must adapt in order to survive, but in this case, that adaptation is based on how we respond to our environment and reciprocally understand how it responds to us. Whether we're aware of this or not, everything we do impacts the world in which we live, and ultimately this synergy reverts to how the world impacts our lives as individuals, as well.

Thoughts on how to participate

Many of us are discouraged from doing anything original by societies that don't value independent critical thinking. Although our skills may be useful at work, many of us stop short from making contributions of our own. That may be a mistake. Now more than ever before, people have opportunities to be creative and share with others worldwide. Digital photography and publishing, along with web pages, blogs, digital books, print on demand, music, and videos have given us media forms that once could only be used by the elite.

The fact that these are available on such a massive scale may seem to detract from their usefulness in attracting public attention, but I think we may be underestimating our potentials. Now that we have the opportunities to do things that once could only have been imagined, people take too much for granted and don't appreciate what they can actually do with inexpensive digital technologies that are available to nearly everyone. Each of us can share.

Being responsive and using feedback

If you enjoyed or found things of interest in this book please leave a review. Your response is appreciated even if it's criticism that foments further debate.

Here's the book link for *The Tao of Systems Thinking*:

https://www.amazon.com/Tao-Systems-Thinking-Exploring-Parallels-ebook/dp/B01A58DXU0

Don't forget that everyone has a creative talent which can be shared with others. I hope you have your own ideas to share about The Tao and Systems Thinking as well. I look forward to your response.

Afterword:

Systems Thinking, like the Tao, is related to how we perceive and respond to our existence in the universe. Although we are limited by how our thinking has been formed as individuals, there's a growing collective understanding of Systems Thinking and world dynamics which can help determine who we are, and what our future could be. While we may not overcome the limitations of our own perceptions, since these are not always right, it may be possible for us to attain a common awareness over time that will help us work to build better societies and live together in greater harmony on this planet.

On the passing of Jay W. Forrester (1918-2016)

Here are a few parting thoughts about a man whose humbleness and curiosity inspired a generation of scientists and thinkers and whose ideas still influence us today. Dr. Jay Forrester pioneered inventions such as magnetic RAM memory for computer technology long before personal computers existed. He founded the field of System Dynamics and helped develop concepts and software for computer simulation modeling that have made a profound impact on our world today. Previously, none of this existed. What he did was intended to help us think more clearly and consider our future.

People who have contributed to Systems Thinking:

Ada Lovelace, Alan Turing, Albert Einstein, Alejandro Jenkins, Alexander Boganov, Alexander von Humbolt, Ann H. Ehrlich, Aristotle, Arthur C. Clarke, Barry Richmond, B.F. Skinner, Benjamin Franklin, Benjamin Hoff, Benoit Mandelbrot, Buckminister Fuller, Candace Perth, Charles Babbage, Charles Darwin, Carl Sagan, Dennis Meadows, Diana Fisher, Donnella Meadows, Dorian Sagan, Draper Kaufmann, Eckhart Tolle, Edward Lorenz, Edwin Abbot, Eric Jantsch, Erwin Schrodinger, Francis Bacon, Francisco Varela, Fritjoff Capra, Garret Hardin, Galileo Galilei, Giordano Bruno, Gregor Mendel, Gregory Bateson, Gottfried Wilhelm Leibniz, Heinz von Foester, Henry David Thoreau, Henry George, Herman Daly, H.G. Wells, Humberto Maturana, Henri Benard, Heraclitus, Immanual Kant, Isaac Asimov, Isaac Newton, Ilya Prigogine, James Clerk Maxwell, James Lovelock, Jay W. Forrester, Jean Piaget, Jeremy Riffkin, John Sterman, John van Neumann, Jorgen Randers, Juan Martin Garcia, Jules Henri Poincare, Jules Verne, Kim Warren, Karl Friedrich Gauss, Lao Tzu, Lees N. Stuntz, Leonardo da Vinci, Leon Festinger, Leonard Euler, Leslie A. Martin, Linda Booth Sweeney, Ludwig Boltzman, Ludwig von Bertanlaffy, Lucia Breierova, Lynn Margulis, Magne Myrtviet, Marie Curie, Margaret Meade, Michael Goodman, Michael J. Radzicki, Nan Lux, Nancy Roberts, Nathan Forrester, Nicolaus Copernicus, Noahm Chomsky, Norbert Weiner, Nikola Tesla, Paul Ehrlich, Peter Senge, Pierre Simon Laplace, Rachel Carson, Rene Descartes,

Richard Alley, Robert Lillienfeld, Ross Ashby, Rudolf Clausius, Scott Adams, Stephanie Albin, Stephen Carlson, Stephen Hawkins, Stafford Beer, Stuart Kauffman, Sun Tsu, Thomas S. Fiddaman, Thomas Malthus, Tycho Brahe, Werner Heisenberg, William Bateson, William Ross Ashby, William Wohlson Behrens Jr., Yoshisuke Ueda, Zeno, and many more.

End notes for Further Reading

The endnotes for this book provide sources with useful online links, but are not formal references with page citations due to the fact that many sources are web pages.

Introductory passages for chapters 1-9 come from "The Wisdoms of Systems Thinking." by M. McCurley (105)

1. Our Universe

1. Edwin Abbot, *Flatland*. (1)

2. Buckminster R. Fuller and E. J. Applewhite, *Synergetics*. (57)

3. Guy Murche, *Music of the Spheres*. (117)

4. Fritjof Capra, *The Web of Life*. (23)

5. Jeremy Rifkin and Ted Howard, *Entropy*. (132)

6. Stephen Hawkins, *A Brief History of Time*. (72)

7. Fritjof Capra, *The Turning Point*. (22)

8. Donella H. Meadows, et al., *The Limits to Growth*. (109)

9. Ludwig Von Bertanlaffy, *General Systems Theory*. (13)

10. Jay W. Forrester, "Counterintuitive Behavior of Social Systems." (49)

11. Naomi Klein, *The Shock Doctrine*. (88)

12. Donella H. Meadows, et. al., *The Limits to Growth*. (109)

2. Balance

1. Donella H. Meadows, et. al., *The Limits to Growth*. (109)

2. Henry George, *Progress and Poverty*. (64)

3. Henry George, Ibid. (64)

4. Naomi Klein, *The Shock Doctrine*. (88)

5. Buckminster R. Fuller, *Synergetics*. (57)

6. Donella H. Meadows, *Thinking in Systems*. (107)

7. Donella H. Meadows, Ibid. (107)

8. Henry David Thoreau, *Walden*. (162)

9. Agnes George de Mille, "Who Was Henry George?" (62)

3. The Great Design

1. B.F. Skinner, *Beyond Freedom and Dignity*. (144)

2. Noam Chomsky, "The Case Against B.F. Skinner." (28)

3. David Foster Wallace, "In his own words." (55)

4. Buckminster R. Fuller, *Synergetics*. (57)

5. John D. Sterman, "All models are wrong." (149)

6. Chris Chatham, "10 Important Differences Between Brains and Computers." (26)

7. Frances Moore Lappé and Frijof. Capra, "Hope is what we become in action." (114)

8. Fritjof Capra, *The Web of Life*. (23)

9. Stephen Hawkins, *A Brief History of Time*. (72)

10. Carl Jung, "The Concept of the Collective Unconscious." (84)

11. Shaikat Hossain, "The Internet as a Tool for Studying the Collective Unconscious." (77)

12. Eckhart Tolle, *A New Earth*. (163)

4. Models and Government

1. "Jay W. Forrester," **MIT Sloan School of Management**. (83)

2. Naomi Klein, *The Shock Doctrine*. (88)

3. "Nuclear Notebook: Global Nuclear Stockpiles," **Natural Resources Defense Council**. (122)

4. "The Cold War (1945-1989)," **CSVE**. (159)

5. "World War II: Facts, information and articles about World War II, 1939-1945," **History Net**. (176)

6. "Relatedness," The Tech Museum of Innovation. (129)

7. Donella H. Meadows, *Thinking in Systems*. (107)

8. John D. Sterman, "Learning from Evidence in a Complex World." (150)

9. Jay W. Forrester, "Learning through System Dynamics as Preparation for the 21st Century." (51)

10. Annie Lowry, "Income Inequality May Take Toll on Growth." (96)

11. Donella H. Meadows, *The Limits to Growth: The 30 Year Update*. (111)

12. N. K. Srinivasan, "Computer Based Modelling and Simulation." (147)

13. Raymond C. Shreckengost, "Dynamic Simulation Models: How Valid Are They?" (142)

14. "The difference between chemical and nuclear energy,"**chemsite. lsrhs.net**. (160)

15. Max Velmans, "Can evolutionary theory explain the existence of consciousness?" (167)

5. Dynamics

1. Jane Cull, "The Circularity of Life." (36)

2. Richard Gaughan, "Solar Power Vs. Fossil Fuels." (60)

3. Jeremy Rifkin, *Entropy*. (132)

4. Peter Andras, "Modeling Living Systems." (6)

5. Brian Fabbri, "Income Inequality and the Limits to Capitalism." (47)

6. Sir John Glubb, "The Fate of Empires." (69)

7. Karim Chichakly, "SMILE: A Common Language for System Dynamics." (27)

8. Jerry Manas, "Complex Systems Have Simple Roots." (102)

9. D.J. Witherspoon, et al., "Genetic Similarities Within and Between Human Populations." (173)

10. Francisco Fernflores, "The Equivalence of Mass and Energy." (48)

11. Peter Tyson, "The Legacy of E = mc2." (165)

12. Key Sun, "Using Taoist Principle of the Unity of Opposites to Explain Conflict and Peace." (151)

13. "Concepts and Frameworks: The Five Learning Disciplines," **Donella Meadows Institute**. (34)

14. Barry Richmond, *An Introduction to Systems Thinking*. Chapter I (131)

15. Dr. Matjaž Mulej, "Systems theory – a worldview and/or a methodology." (116)

16. Elizabeth Baseman, "Political Paradoxes." (10)

17. Randall B. Holcombe, "Political Capitalism." (76)

18. Sun Tzu, *The Art of War*. (152)

19. Simon Black, "These Two Countries are Great Examples of Why Small Governments Work." (17)

20. John Swift, "The Soviet-American Arms Race." (154)

21. Shad Mickelberry, "Cold War Influences on American Culture, Politics, and Economics." (113)

22. Christopher Morris, *The Day They Lost the H-Bomb*. (115)

23. Steven Benner, "Defining Life." (12)

24. Jeremy Rifkin, *Entropy*. (132)

25. Tam Hunt, "The Future of Energy: How to Transition to a Renewable Economy." (80)

26. James Conca, "EROI—A Tool To Predict The Best Energy Mix." (33)

6. Understanding

1. Garret Hardin, "The Tragedy of the Commons." (71)

2. "Expanding Community Rights Over Natural Resources," **Ford Foundation**. (46)

3. Jeffrey D. Sachs, "The Earth provides enough to meet everyone's needs." (135)

4. Naomi Klein, *The Shock Doctrine*. (88)

5. Donella H. Meadows, "Silent Spring and the Population Bomb." (112)

6. Sir John Glubb, "The Fate of Empires." (69)

7. John D. Sterman, "All models are wrong: reflections on becoming a systems scientist." (149)

8. Draper Kauffman, *Systems I.* (85)

9. Jay W. Forrester, *World Dynamics.* (53)

10. William D. Nordhaus, "World Dynamics: Measurement without Data." (120)

11. Magne Myrtveit, "The World Model Controversy." (118)

12. Jay W. Forrester, *World Dynamics.* (53)

13. William D. Nordhaus, "Lethal Model 2: The Limits to Growth Revisited." (119)

14. Brian Hayes, "The Limits to Growth and the limits to computer modeling." (74)

15. Gerald F. Seib, "In Crisis, Opportunity for Obama." (138)

16. Chris Turner, "Sustainability is a process of creative destruction." (164)

17. Kim Warren and Robert Thurlby, "Understanding and Managing the Threat of Disruptive Events to the Critical National Infrastructure." (170)

18. Adam Kowol, "The theory of cognitive dissonance." (89)

19. "Event Oriented Thinkers and System Thinkers," **twink.org**. (45)

20. Jeremy Rifkin, *Entropy.* (132)

21. "Ocean acidification due to increasing atmospheric carbon dioxide," **The Royal Society**. (123)

22. Derek Taylor, "How Much Carbon Dioxide?" (156)

23. David Biello, "How Much Is Too Much?" (14)

24. Mike Gaworecki, "Are We Witnessing The Beginning Of The End For Fossil Fuels?" (61)

25. Michael Taylor, "Summary for Policy Makers: Renewable Power Generation Costs." (157)

26. Matthias Schmelzer and Iris Borowy, "Sustainable Development: the International Struggle over Wealth, Distribution and Limits." (136)

27. Randall B. Holcombe, "Political Capitalism." (76)

28. David A Armor and Shelley E. Taylor, "The Effects of Mindset on Behavior." (7)

29. Howard Wolinsky, "Paths to acceptance. The advancement of scientific knowledge is an uphill struggle against 'accepted wisdom'." (174)

7. Governing Causes

1. "Giordano Bruno," **wikipedia.org**. (66)

2. Tony C. Clark, "Your Perception IS Your Reality." (30)

3. Cordelia Osewa-Ediae, "Systems Thinking: On keeping an open mind and a closed mouth." (124)

4. Henry David Thoreau, *Walden*. (162)

5. Charles Becker, *Progress and Power*. (11)

6. Leslie A. Martin, "Exploring S-Shaped Growth." (103)

7. Eirik Romstad, "Dynamic Efficiency for Stock Pollutants." (133)

8. John Smart, "The Political-Economic Pendulum: The United States Example." (145)

9. Ruth Patterson, "The United States Power Is Declining, and China Is Filling the Vacuum." (125)

10. Nelson Repenning, "Nobody Ever Gets Credit for Fixing Problems that Never Happened." (130)

11. Darell Huff, *How to lie with statistics*. (79)

12. Kambiz E. Maani,"Links Between Systems Thinking and Complex Problem Solving." (97)

13. Malcom Gladwell, *The Tipping Point*. (67)

14. Lucia Breierova, "Generic Structures: Overshoot and Collapse." (18)

15. Karlyn Adams, "The Sources of Innovation and Creativity." (2)

16. Steven A. Benner, "Defining Life." (12)

17. Fritjof Capra, *The Web of Life*. (23)

8. Control

1. Maureen L. Condic, "When Does Human Life Begin?" (35)

2. Roberta Estes, "Human Genetics Revolution Tells Us That Men and Women Are Not the Same." (44)

3. Stephanie Albin, "Generic Structures: First-Order Positive Feedback." (3)

4. DA Southern, "The Battle Within; Why Do We Fight Ourselves – 3 Ways to Balance Our Life Approach." (146)

5. Malcom Gladwell, *The Tipping Point*. (67)

6. Jeanine Prime and Elizabeth Salib, "The Best Leaders Are Humble Leaders." (127)

7. Richard J. Schneider, "Thoreau's Life." (137)

8. "Mahatma Gandhi," **www.history.co.uk** (99)

9. "Martin Luther King Jr.," **biography.com**. (104)

10. "Biography of Nelson Mandela," **The Nelson Mandela Foundation**. (15)

11. "Aung San Suu Kyi," **biography.com**. (9)

12. Sergey A. Vodovozov, "The Gorbachev Era: Perestroika and Glasnost." (168)

13. Ezra F. Vogel, *Deng and the Transformation of China*. (169)

14. "Deng Xiaoping Biography," **biography.com**. (41)

15. Sun Tzu, *The Art of War*. (152)

16. Isaac Asimov, *Foundation*. (8)

17. Sun Tzu, *The Art of War*. (152)

18. Benjamin Franklin, *The Autobiography of Benjamin Franklin*. (56)

19. Richard Paul, "Dominating Egocentrism." (126)

20. Elizabeth Howell, "Challenger: Shuttle Disaster That Changed NASA." (78)

21. James Debner and Larry Jacoby, "Unconscious Perception: Attention, Awareness, and Control." (40)

22. Jessica Davidson, "Evolution of Consciousness: the Dark Side of the Ego." (39)

23. Rory B. Mackay, "Money Is NOT real." (98)

24. David Langford, "Three Faces of Horror." (90)

25. Angela Pritchard, "The Danger of Becoming the Darkness When Fighting the Dark Side." (128)

26. Glenn Llopis, "Leadership Is About Enabling The Full Potential In Others." (94)

9. Wisdom

1. Arlene K. Unger, "Overreacting." (166)

2. James Lovelock, "The Quest for Gaia." (95)

3. Paul Ehrlich, *The Population Bomb*. (43)

4. "World Population Growth," **World Bank**. (175)

5. Dennis Dimick, "As World's Population Booms, Will Its Resources Be Enough for Us?" (42)

6. Charles Darwin, *The Origen of the Species*. (38)

7. Francis Galton, *Hereditary Genius*. (59)

8. Karen Norrgard, "Human testing, the Eugenics Movement, and IRBs." (121)

9. Edwin Black, "The Horrifying American Roots of Nazi Eugenics." (16)

10. Michelle Maiese, "Destructive Escalation." (100)

11. John Sides, "Why does government fail so often?" (143)

12. Peter Senge, "The Necessary Revolution: Creating a Sustainable Future." (141)

13. Brad Waters, "10 Traits of Emotionally Resilient People." (171)

14. Richard Alley, "Abrupt Climate Change." (5)

15. Donella H. Meadows, *Beyond the Limits*. (106)

16. Tony Wilson, "The Dual Nature of Human Behavior." (172)

17. Navid Ghaffarzadegan, et al., "How Small System Dynamics Models Can Help the Public Policy Process." (65)

18. Elisa Teipel, *From Waste to Resource*. (158)

19. James J. Rooney, and Lee N. Vanden Heuvel, "Root Cause Analysis For Beginners." (134)

20. Mike Bundrant, "When you End up with the Opposite of What you Want." (20)

21. Nelson Repenning, "Nobody Ever Gets Credit for Fixing Problems that Never Happened." (130)

22. Robert D. Coleman, "What is Circular Reasoning?" (31)

23. Lauren Steele, "California Is Drying Out and Burning Up." (148)

24. Kim Ann Zimmermann, "Hurricane Katrina: Facts, Damage & Aftermath." (177)

25. Daisaku Ikeda, "Compassion, Wisdom and Courage: Building a Global Society of Peace and Creative Coexistence." (82)

26. Peter Senge, "Navigating Webs of Interdependence." (139)

Tao Te Ching (English translation by Michael McCurley) from Spanish version by Alfonso Colodrón, *Tao Te King* (32)

Bibliography of Links and References

This bibliography contains texts and hyperlinks for further reading.

1-Abbott, Edwin A. *Flatland: Romance of Many Dimensions*. New York: Dover, 1952. Print.

2-Adams, Karlyn. "The Sources of Innovation and Creativity." **National Center on Education and the Economy**. http://www.fpspi.org/ September 2005. Accessed July 15, 2015.
http://www.fpspi.org/pdf/innovcreativity.pdf

3-Albin, Stephanie and Mark Choudhari. "Generic Structures: First-Order Positive Feedback." **Sloan School of Management.** System Dynamics Self Study, Fall 1998 – Spring 1999. March 8, 1996. Accessed June 15, 2015.
http://ocw.mit.edu/courses/sloan-school-of-management/15-988-system-dynamics-self-study-fall-1998-spring-1999/readings/genericpositive.pdf

4-Albin, Stephanie. "Generic Structures: First-Order Negative Feedback." **Sloan School of Management**. System Dynamics Self Study, Fall 1998 – Spring 1999. September 5, 1996. Accessed June 15, 2015.
http://ocw.mit.edu/courses/sloan-school-of-management/15-988-system-dynamics-self-study-fall-1998-spring-1999/readings/genericnegative.pdf

5-Alley, Richard B. "Abrupt Climate Change." *Scientific American*, November, 2004. Accessed May 18, 2015.
http://funnel.sfsu.edu/courses/gm310/articles/SciAm.Abr uptClimateChange.pdf

6-Andras, Peter, "Modeling Living Systems." **School of**

Computing Science, Newcastle University. United Kingdom, 2009. Accessed July 11, 2015. https://www.staff.ncl.ac.uk/peter.andras/PAmodlivECAL200 9.pdf

7-Armor, David A. and Shelley E. Taylor. "The Effects of Mindset on Behavior: Self-Regulation in Deliberative and Implemental Frames of Mind." http://citeseerx.ist.psu.edu/ at UCLA COLLEGE SERIALS/YRL on September 19, 2007. **Society for Personality and Social Psychology, Inc.** Accessed July 12, 2015. http://citeseerx.ist.psu.edu/viewdoc/download?doi=10.1.1.38 5.8036&rep=rep1&type=pdf

8-Asimov, Isaac. *Foundation*. Gnome Press, 1951. Print.

9-"Aung San Suu Kyi." **Bio biography.com**. Accessed April 16, 2015. http://www.biography.com/people/aung-san-suu-kyi-9192617

10-Baseman, Elizabeth. "Political Paradoxes." http://morozshs11ap.weebly.com/ Accessed June 30, 2015. http://morozshs11ap.weebly.com/uploads/1/0/5/5/10558090/political_paradoxes.pdf

11-Becker, Carl L. *Progress and Power.* New York: A.A. Knopf, 1949. Print.

12-Benner, Steven A. "Defining Life." *Astrobiology*, 2010 Dec; 10(10): 1021–1030. **National Institutes of Health**. US **National Library of Medicine**. Accessed July 9, 2015. http://www.ncbi.nlm.nih.gov/pmc/articles/PMC3005285/

13-Bertalanffy, Ludwig Von. *General System Theory; Foundations, Development, Applications.* New York: George Braziller, 1969. Print.

14-Biello, David. "How Much Is Too Much? Estimating Greenhouse Gas Emissions." *Scientific American*, April 29, 2009. Accessed May 7, 2015.

15-"Biography of Nelson Mandela." **The Nelson Mandela Foundation**. https://www.nelsonmandela.org Accessed August 1, 2015.
https://www.nelsonmandela.org/content/page/biography

16-Black, Edwin. "The Horrifying American Roots of Nazi Eugenics." historynewsnetwork.org/ September 2003. (Originally published in the *San Francisco Chronicle*) Accessed July 11, 2015.
http://historynewsnetwork.org/article/1796

17-Black, Simon. "These Two Countries are Great Examples of Why Small Governments Work." **Sovereign Man.** *Business Insider*, April 2, 2012. Accessed July 17, 2015.
http://www.businessinsider.com/democracy-looks-great-on-paper-until-2012-4

18-Breierova, Lucia. "Generic Structures: Overshoot and Collapse." mit.edu/courses. **Sloan School of Management**. System Dynamics Self Study, Fall 1998 – Spring 1999. July 21, 1997. Accessed June 15, 2015.
http://ocw.mit.edu/courses/sloan-school-of-management/15-988-system-dynamics-self-study-fall-1998-spring-1999/readings/generic3.pdf

19-Bremner, Jason. "Population, poverty, environment, and climate dynamics in the developing world."
http://geog.ucsb.edu/~carr/wordpress/wp-.content/uploads/2012/04/Bremneretal_IntEnvRev_201012.pdf

20-Bundrant, Mike. "When you End up with the Opposite of What you Want." NLP Discoveries. *Psych Central*, May 2013. Accessed July 13, 2015.
http://blogs.psychcentral.com/nlp/2013/05/self-sabotage-opposite-of-what-you-want/

21-Capra, Fritjof. *The Tao of Physics: An Exploration of the Parallels between Modern Physics and Eastern Mysticism.* Berkeley: Shambhala, 1975. Print.

22-Capra, Fritjof. *The Turning Point: Science, Society and the Rising Culture.* Simon and Schuster: New York, 1982. Print.

23-Capra, Fritjof. *The Web of Life: A new scientific understanding of living systems.* New York: Anchor Books, 1996. Print.

24-Carson, Rachel and Louis Darling. *Silent Spring.* Boston: Houghton Mifflin, 1962. Print.

25-Carson, Rachel. *Silent Spring.* 1962. http://library.uniteddiversity.coop/More_Books_and_Reports /Silent_Spring-Rachel_Carson-1962.pdf

26-Chatham, Chris. "10 Important Differences Between Brains and Computers." March 27, 2007. Developing Intelligence, in **Science Blogs**. Accessed May 7, 2015. http://scienceblogs.com/developingintelligence/2007/03/27/ why-the-brain-is-not-like-a-co/

27-Chichakly, Karim. "SMILE: A Common Language for System Dynamics." **Information Systems SIG of the Systems Dynamics Society**. Language Subcommittee, June 7, 2013. http://www.iseesystems.com/community/support/SMILEv4. pdf

28-Chomsky, Noam. "The Case Against B.F. Skinner." *The New York Review of Books*, December 30, 1971. Accessed July 11, 2015. http://www.chomsky.info/articles/19711230.htm

29-Chung, Celeste V. "Generic Structures in Oscillating Systems." mit.edu/courses. **Sloan School of Management**. System Dynamics Self Study, Fall 1998 – Spring 1999.

http://ocw.mit.edu/courses/sloan-school-of-management/15-988-system-dynamics-self-study-fall-1998-spring-1999/readings/oscillating.pdf

30-Clark, Tony C. "Your Perception IS Your Reality." http://www.lifehack.org/ **Lifestyle**. Accessed July 6, 2015. http://www.lifehack.org/articles/lifestyle/your-perception-is-your-reality.html

31-Coleman, Robert D. "What is Circular Reasoning?" http://www.numeraire.com/ 2006. July 12, 2015. http://www.numeraire.com/download/WhatIsCircularReasoning.pdf

32-Colodrón, Alfonso. Translation of *Tao Te King*. (de la traducción de John C. H. Wu) Editorial Edaf: Madrid, 1993. Print.

33-Conca, James. "EROI -- A Tool To Predict The Best Energy Mix." **http://www.forbes.com** February 11, 2015. Accessed June 29, 2015. http://www.forbes.com/sites/jamesconca/2015/02/11/eroi-a-tool-to-predict-the-best-energy-mix/

34-"Concepts and Frameworks: The Five Learning Disciplines." Systems Thinking Resources. **Donella Meadows Institute**. Accessed Juy 16, 2015. http://www.donellameadows.org/systems-thinking-resources/

35-Condic, Maureen L. "When Does Human Life Begin? A Scientific Perspective." **http://bdfund.org/** http://bdfund.org/wordpress/wp-content/uploads/2012/06/wi_whitepaper_life_print.pdf

36-Cull, Jane. "The Circularity of Life." *The Ecologist*, November 4, 2014. Accessed June 24, 2015. http://www.theecologist.org/blogs_and_comments/Blogs/2560462/the_circularity_of_life.html

37-Darwin, Charles. *On the Origin of Species by Means of Natural Selection, or the Preservation of Favored Races in the Struggle for Life*. 1ˢᵗ ed. London: John Murray, 1859. Print.

38-Darwin, Charles. *The Origin of the Species*. 1873 version. https://www.andrew.cmu.edu/user/jksadegh/A%20Good%20 Atheist%20Secularist%20Skeptical%20Book%20Collection/ Charles%20Darwin%20- %20The%20Origin%20of%20Species%20- %206th%20Edition.pdf

39-Davidson, Jessica. "Evolution of Consciousness: the Dark Side of the Ego." Jessica Davidson Right Mindfulness, Write Thought. **http://jessicadavidson.co.uk/** April 13, 2015. Accessed July 11, 2015. http://jessicadavidson.co.uk/2015/04/13/evolution-of-consciousness-the-dark-side-of-the-ego/

40-Debner, James A. and Larry L. Jacoby. "Unconscious Perception: Attention, Awareness, and Control." *Journal of Experimental Psychology*: Learning, Memory, and Cognition. American Psychological Association, 1994. Accessed July 10, 2015. http://citeseerx.ist.psu.edu/viewdoc/download?doi=10.1.1.41 2.4083&rep=rep1&type=pdf

41-"Deng Xiaoping Biography." **Bio. biography.com**. http://www.biography.com/people/deng-xiaoping-9271644

42-Dimick, Dennis. "As World's Population Booms, Will Its Resources Be Enough for Us?" *National Geographic*, September 21, 2014. Accessed April 15, 2015. http://news.nationalgeographic.com/news/2014/09/140920-population-11billion-demographics-anthropocene/

43-Ehrlich, Paul R. *The Population Bomb*. New York: Ballantine, 1968. Print.

44-Estes, Roberta. "Human Genetics Revolution Tells Us That Men and Women Are Not the Same." **DNAeXplained – Genetic Genealogy**, Oct. 24, 2013. Accessed July 22, 2015. http://dna-explained.com/2013/10/24/human-genetics-revolution-tells-us-that-men-and-women-are-not-the-same/

45-"Event Oriented Thinkers and System Thinkers." **Systems Thinking. twink.org.** http://www.thwink.org Accessed July 15, 2015. http://www.thwink.org/sustain/glossary/SystemsThinking.htm

46-"Expanding Community Rights Over Natural Resources." **Ford Foundation. http://www.fordfoundation.org/** September 2010. Accessed July 16, 2015. http://www.fordfoundation.org/pdfs/issues/community-rights-initiative-overview.pdf

47-Fabbri, Brian. "Income Inequality and the Limits to Capitalism: The Haves and the Have-Nots." **National University of Singapore**, 2012. Accessed July 16, 2015. http://bschool.nus.edu.sg/Portals/0/images/CAMRI/Thought%20Leadership/Income%20Inequality%20and%20the%20Limits%20to%20Capitalism-The%20Haves%20and%20HaveNots_October%202012_Brian%20Fabbri.pdf

48-Fernflores, Francisco, "The Equivalence of Mass and Energy." *The Stanford Encyclopedia of Philosophy*, (Spring 2012 Edition), Edward N. Zalta (ed.) Accessed July 11, 2015 http://plato.stanford.edu/entries/equivME/ http://plato.stanford.edu/archives/spr2012/entries/equivME/

49-Forrester, Jay W. "Counterintuitive Behavior of Social Systems." Accessed April 15, 2015. http://www.constitution.org/ps/cbss.pdf

50-Forrester, Jay W. "Industrial Dynamics-After the First Decade." Accessed April 15, 2015.

http://www.sfu.ca/~vdabbagh/Forrester68.pdf

51-Forrester, Jay W. "Learning through System Dynamics as Preparation for the 21st Century." **Sloan School of Management**. System Dynamics Self Study, Fall 1998 – Spring 1999. PDF. Accessed June 15, 2015. http://ocw.mit.edu/courses/sloan-school-of-management/15-988-system-dynamics-self-study-fall-1998-spring-1999/readings/learning2.pdf

52-Forrester, Jay Wright. *Principles of Systems*. Portland: Productivity, 1971. Print.

53-Forrester, Jay W. *World Dynamics*. Cambridge, MA: Wright-Allen, 1971. Print.

54-Forrester, Jay W. *Urban Dynamics*. Cambridge, MA: M.I.T., 1969. Print.

55-Foster Wallace, David. "David Foster Wallace, in his own words." In Memoriam, September 19, 2008. *The Economist*. Intelligent Life. Accessed July 17, 2015. http://moreintelligentlife.com/story/david-foster-wallace-in-his-own-words

56-Franklin, Benjamin. *The Autobiography of Benjamin Franklin.* **eBooks@Adelaide.** The University of Adelaide, Australia. First published 1791, 1793. Updated December 17, 2014. Accessed July 11, 2015. https://ebooks.adelaide.edu.au/f/franklin/benjamin/autobiography/complete.html

57-Fuller, R. Buckminster and E. J. Applewhite. *Synergetics: Explorations in the Geometry of Thinking*. New York: Macmillan, 1975. Print.

58-Galton, Francis. *Hereditary Genius*. London: Macmillan, 1869. Print.

59-Galton, Francis. *Hereditary Genius*. 1869 http://www.mugu.com/galton/books/hereditary-genius/galton-1869-Hereditary_Genius.pdf

60-Gaughan, Richard. "Solar Power Vs. Fossil Fuels." http://education.seattlepi.com **Demand Media**. Accessed June 29, 2015. http://education.seattlepi.com/solar-power-vs-fossil-fuels-3937.html

61-Gaworecki, Mike. "Are We Witnessing The Beginning Of The End For Fossil Fuels?" **DESMOG CLEARING THE PR POLLUTION THAT CLOUDS CLIMATE SCIENCE**, May 17, 2015. Accessed July 8, 2015. http://www.desmogblog.com/2015/05/17/are-we-witnessing-beginning-end-fossil-fuels

62-George de Mille, Agnes. "Who Was Henry George?" 1979. http://www.henrygeorge.org/whowashg.htm

63-George, Henry. *Progress and Poverty*. Garden City, NY: Doubleday, Page & Co., 1879. Print.

64-George, Henry. *Progress and Poverty*. https://mises.org Accessed August 1, 2015. https://mises.org/sites/default/files/Progress%20and%20Poverty_3.pdf

65-Ghaffarzadegan, Navid, John Lyneis, and George P. Richardson. "How Small System Dynamics Models Can Help the Public Policy Process." Accessed July 13, 2015. http://www.albany.edu/~gpr/SmallModels.pdf

66-"Giordano Bruno." **Wikipedia**. **wikipedia.org** Accessed July 8, 2015. https://en.wikipedia.org/wiki/Giordano_Bruno

67-Gladwell, Malcolm. *The Tipping Point: How Little Things Can Make a Big Difference*. Boston: Little, Brown, 2000. Print.

68-Glick, Marc and Terry Duhon. "Generic Structures: S-Shaped Growth." **Sloan School of Management**. mit.edu/courses System Dynamics Self Study, Fall 1998 – Spring 1999.

http://ocw.mit.edu/courses/sloan-school-of-management/15-988-system-dynamics-self-study-fall-1998-spring-1999/readings/growth.pdf

69-Glubb, Sir John. "The Fate of Empires." **www.newworldeconomics.com** Accessed June, 27, 2015. http://www.newworldeconomics.com/archives/2014/092814 _files/TheFateofEmpiresbySirJohnGlubb.pdf

70-Hall, William P. "Knowledge and Diversity in Complex Systems: Emergence and Growth of Knowledge and Diversity in Hierarchically Complex Living Systems." **Australian Centre for Science, Innovation and Society. University of Melbourne**, Vic., Australia, 2006. Accessed June 24, 2015. file:///C:/Users/Guest/Downloads/SSRN-id1758090.pdf http://papers.ssrn.com/sol3/papers.cfm?abstract_id=1758090

71-Hardin, Garret. "The Tragedy of the Commons." 1968. http://www.geo.mtu.edu/~asmayer/rural_sustain/governance/ Hardin%201968.pdf

72-Hawking, Stephen. *A Brief History of Time: From the Big Bang to Black Holes*. Toronto: Bantam, 1988. Print.

73-Hawking, Stephen. **A Brief History of Time**. 1988. Accessed April 18, 2015. http://www.fisica.net/relatividade/stephen_hawking_a_brief_ history_of_time.pdf

74-Hayes, Brian. "The Limits to Growth and the limits to computer modeling." Computation and the Human Predicament. *American Scientist*. Accessed July 17, 2015. http://www.americanscientist.org/libraries/documents/20124 91358139046-2012-05Hayes.pdf

75-Hoff, Benjamin. *The Tao of Pooh*. New York, NY: Penguin, 1983. Print.

76-Holcombe, Randall B. "Political Capitalism." *Cato Journal.* **Cato Institute**: Winter 2015. Accessed July 17, 2015.

http://object.cato.org/sites/cato.org/files/serials/files/cato-journal/2015/2/cj-v35n1-2.pdf

77-Hossain, Shaikat. "The Internet as a Tool for Studying the Collective Unconscious." *Jung Journal*: Culture & Psyche 6:2 / Spring 2012. Accessed July 26, 2015. http://www.utdallas.edu/~sxh096020/JUNG.pdf

78-Howell, Elizabeth. "Challenger: Shuttle Disaster That Changed NASA." **Space.com**. October 16, 2012. Accessed July 11, 2015 http://www.space.com/18084-space-shuttle-challenger.html

79-Huff, Darrell. *How to Lie with Statistics*. England: Clays Ltd, St Ives plc. Penguin Books, 1971. Print.

80-Hunt, Tam. "The Future of Energy: How to Transition to a Renewable Economy." **greentechmedia**. http://www.greentechmedia.com/ October 21, 2014. Accessed July 15, 2015. http://www.greentechmedia.com/articles/read/the-future-of-energy-how-to-transition-to-a-renewable-economy

81-Isaacs, Julia and Isabel Sawhill. "Reaching for the Prize: The Limits on Economic Mobility." 2008. **Brookings Institute**. Accessed July 17, 2015. http://www.brookings.edu/~/media/research/files/papers/200 8/10/winter-economic-mobility-isaacs-sawhill/winter_economic_mobility_isaacs_sawhill.pdf

82-Ikeda, Daisaku. "Compassion, Wisdom and Courage: Building a Global Society of Peace and Creative Coexistence." **http://www.sgi.org/ Soka Gakkai International**, January 26, 2013. Accessed July 13, 2015. http://www.sgi.org/assets/pdf/peaceproposal2013.pdf

83-"Jay W. Forrester." **MIT Sloan School of Management**. Accessed April 19, 2015. http://clexchange.org/ftp/documents/system-dynamics/SD2011-01JayWForresterBio.pdf

84-Jung, Carl. "The Concept of the Collective Unconscious."
Selection from ***Understanding Dreams***.
http://bahaistudies.net/ Accessed August 1, 2015.
http://bahaistudies.net/asma/The-Concept-of-the-Collective-Unconscious.pdf

85-Kauffman, Draper L. ***Systems One: An Introduction to Systems Thinking***. Minneapolis: Future Systems, 1980. Print.

86-Kauffman, Draper. ***Systems one: An introduction to systems thinking***. 1980.
http://www.academia.edu/3317732/Systems_one_An_introd uction_to_systems_thinking

87-Khazan, Olga. "Can we fight poverty by ending extreme wealth?" *The Washington Post*, 2013.
http://www.washingtonpost.com/blogs/worldviews/wp/2013/ 01/20/oxfam-poverty-income-inequality/

88-Klein, Naomi. *The Shock Doctrine: The Rise of Disaster Capitalism*. New York: Metropolitan/Henry Holt, 2007. Print.

89-Kowol, Adam. "The theory of cognitive dissonance."
http://works.adamkowol.info/ Accessed July 26, 2015.
http://works.adamkowol.info/Festinger.pdf

90-Langford, David. "Three Faces of Horror." **ansible.uk**
http://ansible.uk/writing (commissioned for the Penguin UK website in 1999 and long since removed from view) Accessed July 11, 2015.
http://ansible.uk/writing/3horror.html

91-Lao Tzu, *Tao Te Ching*. Circa 500 B.C. [multiple versions and translations are available]

92-Lao Tzu, *Tao Te Ching*. Translated by Stephen Mitchell. **brooklyn.cuny.edu** Updated July 20, 1995. Accessed July 22, 2015.
http://acc6.its.brooklyn.cuny.edu/~phalsall/texts/taote-v3.html

93-Li, Anson and Kambiz Maani. "Dynamic Decision-Making, Learning and Mental Models." **systemdynamics.org**. Accessed June 16, 2015. http://www.systemdynamics.org/conferences/2011/proceed/papers/P1189.pdf

94-Llopis, Glenn. "Leadership Is About Enabling The Full Potential In Others." *Forbes.* http://www.forbes.com/ July 29, 2014. Accessed August 1, 2015. http://www.forbes.com/sites/glennllopis/2014/07/29/leadership-is-about-enabling-the-full-potential-in-others/

95-Lovelock, James and Sidney Epton. "The Quest for Gaia." *New Scientist,* February 6, 1975. Accessed June 30, 2015. https://books.google.co.uk/books?id=pnV6UYEkU4YC&printsec=frontcover&source=gbs_ge_summary_r&hl=en#v=onepage&q&f=false

96-Lowrey, Annie. "Income Inequality May Take Toll on Growth." *The New York Times,* October 16, 2012. http://www.nytimes.com/2012/10/17/business/economy/income-inequality-may-take-toll-on-growth.html

97-Maani, Kambiz E. and Vandana Maharaj. "Links Between Systems Thinking and Complex Problem Solving - Further Evidence." **Department of Management Science and Information Systems. School of Business and Economics. The University of Auckland,** New Zealand. Accessed July 11, 2015. http://citeseerx.ist.psu.edu/viewdoc/download?doi=10.1.1.379.6492&rep=rep1&type=pdf

98-Mackay, Rory B. "Money is NOT real." **Beyond the Dream**. http://beyondthedream.co.uk/ July 7, 2013. Accessed July 12, 2015. http://beyondthedream.co.uk/2013/07/07/money-is-not-real/

99-"Mahatma Gandhi." **history.co.uk**. Accessed April 15, 2015. http://www.history.co.uk/biographies/mahatma-gandhi

100-Maiese, Michelle. "Destructive Escalation." **Beyond Intractability**. Eds. Guy Burgess and Heidi Burgess. **Conflict Information Consortium**. University of Colorado, September 2003.
http://www.beyondintractability.org Accessed July 13, 2015.
http://www.beyondintractability.org/essay/escalation

101-Malthus, Thomas. *An Essay on the Principle of Population.* 1798.
http://www.esp.org/books/malthus/population/malthus.pdf

102-Manas, Jerry. "Complex Systems Have Simple Roots." Official Website of Bestselling Author Jerry Manas. October 30, 2009. Accessed July 9, 2015.
http://jerrymanas.com/2009/10/30/complex-systems-have-simple-roots/

103-Martin, Leslie A. "Exploring S-Shaped Growth." **Sloan School of Management**. mit.edu/courses, System Dynamics Self Study, Fall 1998 – Spring 1999. October 3, 1996. Accessed June 15, 2015.
http://ocw.mit.edu/courses/sloan-school-of-management/15-988-system-dynamics-self-study-fall-1998-spring-1999/readings/exploring.pdf

104-"Martin Luther King Jr." **Bio biography.com**. Accessed April 16, 2015.
http://www.biography.com/people/martin-luther-king-jr-9365086

105-McCurley, Michael. "The Wisdoms of Systems Thinking." **scribd.com** 2011, 2015. Accessed August 16, 2015.
https://www.scribd.com/doc/274669640/The-Wisdoms-of-Systems-Thinking

106-Meadows, Donella H. and J. Randers. *Beyond the Limits*, Chelsea Green Publishing Company, White River Junction VT, 1992. Print.

107-Meadows, Donella H. and Diana Wright. *Thinking in Systems: A Primer*. White River Junction, VT: Chelsea Green Pub., 2008. Print.

108-Meadows, Donella H. *Thinking in Systems*. Accessed April 15, 2015.
http://ir.nmu.org.ua/bitstream/handle/123456789/129200/2ee 4a14a158e824b867e07ad95005643.pdf?sequence=1

109-Meadows, Donella H., Dennis Meadows, Jorgen Randers, and William W. Behrens III. *The Limits to Growth: a Report for the Club of Rome's Project on the Predicament of Mankind*. New York: Universe, 1972. Print.

110-Meadows, Donella, et al. *The Limits to Growth*. Accessed April 15, 2015
http://www.donellameadows.org/wp-content/userfiles/Limits-to-Growth-digital-scan-version.pdf

111-Meadows, Donella H., et al. *The Limits to Growth: The 30-Year Update*. Vermont: Chelsea Green Publishing Company, 2004. Print.

112-Meadows, Donella. "Silent Spring and The Population Bomb—Can Books or Lives Make Any Difference?" 1988.
http://www.donellameadows.org/archives/silent-spring-and-the-population-bomb-can-books-or-lives-make-any-difference/

113-Mickelberry, Shad. "Cold War Influences on American Culture, Politics, and Economics." Shad's Blog Adventures and Random Thoughts. December, 2009. Accessed July 9, 2015.
https://tradshad.wordpress.com/writings/cold-war-influences-on-american-culture-politics-and-economics/

114-Moore Lappé, Frances and Fritjof Capra. "Hope is what we become in action." **Center for Ecoliteracy**, June, 2013.
http://www.ecoliteracy.org/essays/hope-what-we-become-action-frances-moore-lappe-and-fritjof-capra-conversation

115-Morris, Christopher. *The Day they Lost the H-Bomb*. New York: Coward McCann, 1966. Print.
116-Mulej, Dr. Matjaž. "Systems theory – a worldview and/or a methodology." Ross Ashby Memorial Lecture of the IFSR. **International Federation for Systems Research**, 2006. Accessed July 9, 2015.
http://www.ifsr.org/wp-content/uploads/2013/04/Mulej_Ashby_20060115-2006.pdf
117-Murchie, Guy. *Music of the Spheres*. Boston: Houghton Mifflin, 1961. Print.
118-Myrtveit, Magne. "The World Model Controversy." Accessed April 15, 2015.
http://www.folk.uib.no/sinem/WPSD/WPSD1.05WorldContr oversy.pdf
119-Nordhaus, William D. "Lethal Model 2: The Limits to Growth Revisited." **www.brookings.edu**
http://www.brookings.edu/~/media/Projects/BPEA/1992-2/1992b_bpea_nordhaus_stavins_weitzman.PDF
120-Nordhaus, William D. "World Dynamics: Measurement Without Data". *The Economic Journal*, Vol. 83, No. 332. (Dec., 1973), pp. 1156-1183. Accessed April 15, 2015.
http://www.econ.yale.edu/~nordhaus/homepage/worlddynam ics.pdf
121-Norrgard, Karen. "Human testing, the Eugenics Movement, and IRBs." *Nature Education*, 2008.
http://www.nature.com/scitable/topicpage/human-testing-the-eugenics-movement-and-irbs-724
122-"Nuclear Notebook: Global Nuclear Stockpiles." **Natural Resources Defense Council**, 2006. Accessed April 19, 2015.
http://csis.org/files/media/csis/pubs/poni/global_nuclear_stoc kpiles.pdf

123-"Ocean acidification due to increasing atmospheric carbon dioxide." **The Royal Society.** www.royalsoc.ac.uk July, 2005. Accessed August 1, 2015. http://www.us-ocb.org/publications/Royal_Soc_OA.pdf

124-Osewa-Ediae, Cordelia. "Systems Thinking: On keeping an open mind and a closed mouth." **Clore Social Leadership Programme**, June 8, 2015. Accessed July 12, 2015. http://www.cloresocialleadership.org.uk/Cordelia-on-Systems-Thinking

125-Patterson, Ruth. "The United States Power Is Declining, and China Is Filling the Vacuum." **Economy in Crisis.** http://economyincrisis.org/ June 15, 2015, Accessed July 11, 2015. http://economyincrisis.org/content/36291

126-Paul, Richard and Linda Elder. "Dominating Egocentrism." From *Critical Thinking: Tools for Taking Charge of Your Professional and Personal Life* **Financial Times Press**, August 29, 2013. Accessed July 12, 2015. https://books.google.co.cr/books?id=tcuEAAAAQBAJ&pg=PA217&lpg=PA217&dq=dominating+egocentrism,+Richarl+Paul&source=bl&ots=jjCRHhSXWH&sig=5dRp_vHJarOAR2pLZAFHcwP3spM&hl=en&sa=X&ei=3yKjVajfGcHLogSrjaW4BQ&redir_esc=y#v=onepage&q=dominating%20egocentrism%2C%20Richarl%20Paul&f=false

127-Prime, Jeanine and Elizabeth Salib. "The Best Leaders Are Humble Leaders." Leadership. *Harvard Business Review*. https://hbr.org/ May 12, 2014. Accessed July 14, 2015. https://hbr.org/2014/05/the-best-leaders-are-humble-leaders/

128-Pritchard, Angela. "The Danger of Becoming the Darkness When Fighting the Dark Side." http://belsebuub.com/ February 8, 2014. Accessed July 11, 2015. http://belsebuub.com/the-danger-of-becoming-the-darkness-when-fighting-the-dark-side

129-"Relatedness." **The Tech Museum of Innovation**. Stanford at the Tech: Understanding Genetics. Accessed June 16, 2015. http://genetics.thetech.org/ask/ask38

130-Repenning, Nelson and John Sterman. "Nobody Ever Gets Credit for Fixing Problems that Never Happened." 2001. Accessed April 16, 2015. http://jsterman.scripts.mit.edu/docs/Repenning-2001-NobodyEverGetsCredit.pdf

131-Richmond, Barry. *An Introduction to Systems Thinking.* Chapter I. **isee systems** Accessed July 14, 2015. http://www.iseesystems.com/resources/Articles/StellaIST(ch 1).pdf

132-Rifkin, Jeremy and Ted Howard. *Entropy: A New World View.* New York: Viking, 1980. Print.

133-Romstad, Eirik. "Dynamic Efficiency for Stock Pollutants." **School of Economics and Business. Norwegian University of Life Sciences**. https://athene.nmbu.no https://athene.nmbu.no/emner/pub/ECN371/notes/dynamic.pdf

134-Rooney, James J. and Lee N. Vanden Heuvel. "Root Cause Analysis For Beginners." **Quality Basics**, https://servicelink.pinnacol.com Accessed July 11, 2015. https://servicelink.pinnacol.com/pinnacol_docs/lp/cdrom_web/safety/management/accident_investigation/Root_Cause.pdf

135-Sachs, Jeffrey D. "The Earth provides enough to meet everyone's needs." **The National—Opinion**, March 2, 2011. http://www.thenational.ae/ March 2, 2011. Accessed July 15, 2015. http://www.thenational.ae/thenationalconversation/comment/the-earth-provides-enough-to-meet-everyones-needs#full

136-Schmelzer, Matthias and Iris Borowy. "'Sustainable Development': the International Struggle over Wealth, Distribution and Limits." July 17, 2015.

http://co-munity.net/system/files/3267.pdf

137-Schneider, Richard J. "Thoreau's Life." **The Thoreau Society**. Accessed July 8, 2015. http://www.thoreausociety.org/life-legacy

138-Seib, Gerald F. "In Crisis, Opportunity for Obama." *The Wall Street Journal*, Nov. 21, 2008. Accessed July 12, 2015. http://www.wsj.com/articles/SB122721278056345271

139-Senge, Peter. "Navigating Webs of Interdependence." What Is Systems Thinking? – Peter Senge Explains Systems Thinking Approach And Principles. **mutualresponsibility.org** Accessed August 1, 2015. http://www.mutualresponsibility.org/science/what-is-systems-thinking-peter-senge-explains-systems-thinking-approach-and-principles

140-Senge, Peter. *The fifth discipline: The art and practice of the learning organization*. New York: Doubleday/Currency, 1990. Print.

141-Senge, Peter and Bryan Smith. "The Necessary Revolution: Creating a Sustainable Future." **http://changethis.com** August 6, 2008. Accessed July 11, 2015 http://changethis.com/manifesto/49.01.NecessaryRevolution/pdf/49.01.NecessaryRevolution.pdf

142-Shreckengost, Raymond C., "Dynamic Simulation Models: How Valid Are They?" **MIT Sloan School of Management**. System Dynamics Self Study, Fall 1998–Spring 1999." Accessed June 15, 2015. http://ocw.mit.edu/courses/sloan-school-of-management/15-988-system-dynamics-self-study-fall-1998-spring-1999/readings/dynamic.pdf

143-Sides, John. "Why does government fail so often?" Monkey Page *The Washington Post*, May 29, 2014. Accessed July 10, 2015.

http://www.washingtonpost.com/blogs/monkey-cage/wp/2014/05/29/why-does-government-fail-so-often/

144-Skinner, Burrhus Frederic. *Beyond Freedom and Dignity*: New York: A.A. Knopf, 1971. Print.

145-Smart, John. "The Political-Economic Pendulum: The United States Example." Acceleration Watch. http://accelerationwatch.com/ (*From Thinking Creatively in Turbulent Times*, Howard Didsbury, ed., **World Future Society**, 2004) Accessed July 11, 2015. http://accelerationwatch.com/articles/pe_pendulum.html#pendulum

146-Southern, DA "The Battle Within; Why Do We Fight Ourselves – 3 Ways to Balance Our Life Approach." **Direct Your Own Life**. **https://directyourownlife**.wordpress.com March 6, 2012. Accessed July 14, 2015. https://directyourownlife.wordpress.com/2012/03/06/the-battle-within-why-do-we-fight-ourselves-3-ways-to-balance-our-life-approach/

147-Srinivasan, N. K. "Computer Based Modelling and Simulation: Modelling Deterministic Systems." *Resonance*, March, 2001. Accessed June 26, 2015. http://www.ias.ac.in/resonance/Volumes/06/03/0046-0054.pdf

148-Steele, Lauren. "California Is Drying Out and Burning Up." **News from the Field**. http://www.outsideonline.com/ August 19, 2014. Accessed June 30, 2015. http://www.outsideonline.com/1805031/california-drying-out-and-burning

149-Sterman, John D. "All models are wrong: reflections on becoming a systems scientist." **Jay Wright Forrester Prize Lecture**, 2002. http://web.mit.edu/jsterman/www/All_Models_Are_Wrong_%28SDR%29.pdf

150-Sterman, John D. "Learning from Evidence in a Complex World." *American Journal of Public Health*: 2006 March; 96(3): 505–514. **US National Library of Medicine. National Institute of Health.** http://www.ncbi.nlm.nih.gov/pmc/articles/PMC1470513/

151-Sun, Key."Using Taoist Principle of the Unity of Opposites to Explain Conflict and Peace." *The Humanistic Psychologist*, 37: 271–286, 2009. Accessed July 26, 2015. https://www.psychologytoday.com/sites/default/files/attachments/35629/using-taoist-principle-the-unity-opposites.pdf

152-Sun Tzu, *The Art of War*, the Oldest Military Treatise in the World. Translated by Lionel Giles. Champaign, Ill.: **Project Gutenberg**, circa 6ᵗʰ century B.C. Accessed August 1, 2015. http://www.gutenberg.org/files/132/132.txt

153-Sun Tzu. *The Art of War*. Circa 6ᵗʰ century B.C. http://www.puppetpress.com/classics/ArtofWarbySunTzu.pdf

154-Swift, John. "The Soviet-American Arms Race." (First published in *History Review* Issue 63, March 2009. *History Today*.) http://www.historytoday.com/ Accessed July, 9, 2015. http://www.historytoday.com/john-swift/soviet-american-arms-race

155-Tabasum. "Why Flow of Materials in Ecosystem is Cyclic but the Flow of Energy is Unidirectional?" **www.preservearticles.com** Accessed June 24, 2015. http://www.preservearticles.com/2012010219583/why-flow-of-materials-in-ecosystem-is-cyclic-but-flow-of-energy-is-unidirectional.html

156-Taylor, Derek. "How Much Carbon Dioxide?" Solutions to Climate Change. **http://decarboni.se/** November 11, 2011. Accessed July 26, 2015 http://decarboni.se/insights/how-much-carbon-dioxide

157-Taylor, Michael. "Summary for Policy Makers: Renewable Power Generation Costs." IRENA **International Revewable Energy Agency.** www.irena.org Bonn, Germany: Innovation and Technology Centre, November 2012. Accessed July 9, 2015.
https://www.irena.org/DocumentDownloads/Publications/Re newable_Power_Generation_Costs.pdf

158-Teipel, Elisa. "From Waste to Resource: A Systems-Based Approach to Sustainable Community Development through Equitable Enterprise and Agriculturally Derived Polymeric Composites." **University of Colorado, Boulder CU Scholar,** Summer 7- 17-2014. PDF. Accessed June 17, 2015.
file:///C:/Users/Guest/Documents/From%20,   ,
,Waste%20to%20Resource-%20A%20Systems-Based%20Approach%20to%20Sustainable%20C.pdf
http://scholar.colorado.edu/cven_gradetds/9/

159-"The Cold War (1945–1989) — Full text." **Centre Virtuel de la Connaissance sur l'Europe** (CVCE) Last updated 03/07/2014. Accessed April 19, 2015.
http://www.cvce.eu/obj/the_cold_war_1945_1989_full_text-en-6dfe06ed-4790-48a4-8968-855e90593185.html
http://www.cvce.eu/content/publication/2011/11/21/6dfe06e d-4790-48a4-8968-855e90593185/publishable_en.pdf

160-"The difference between chemical and nuclear energy." Chemistry 2. **lsrhs.net**
http://chemsite.lsrhs.net/Nuclear/chemNucDifference.html

161-Thoreau, Henry David. *Walden: Or, Life in the Woods.* Boston, MA: Ticknor & Fields, 1854. Print.

162-Thoreau, Henry David. *Walden.* 1854. Online version. Accessed April, 19, 2015.
http://www.eldritchpress.org/walden5.pdf

163-Tolle, Eckhart. *A New Earth: Awakening to Your Life's Purpose*. New York: Plume, 2006. Print.

164-Turner, Chris. "Sustainability is a process of creative destruction." **mother nature network** http://www.mnn.com/ November 4, 2011. Accessed July 16, 2015. http://www.mnn.com/green-tech/research-innovations/blogs/sustainability-is-a-process-of-creative-destruction

165-Tyson, Peter "The Legacy of E = mc2." **Nova** Physics + Math. http://www.pbs.org/ Accessed June 30, 2015. http://www.pbs.org/wgbh/nova/physics/legacy-of-e-equals-mc2.html

166-Unger, Arlene K. "Overreacting." **www.realpsychsolutions.com** 2010. Accessed July 18, 2015. https://www.realpsychsolutions.com/uploads/overreacting-Unger-rps.pdf

167-Velmans. Max. "Can evolutionary theory explain the existence of consciousness?" A Review of Humphrey, N. (2010) Soul Dust: The Magic of Consciousness. London: Quercus. Reviewed by Max Velmans, Goldsmiths, University of London. *Journal of Consciousness Studies*, Volume 18, No.11-12 (2011), pp. 243-254. http://cogprints.org/7750/1/Humphrey%20review%20final.pdf

168-Vodovozov, Sergey Arsentyevich. "The Gorbachev Era: Perestroika and Glasnost." **www.britanica.com** http://www.britannica.com/EBchecked/topic/513251/Russia/38564/The-Gorbachev-era-perestroika-and-glasnost

169-Vogel, Ezra F. *Deng Xiaopin and the Transformation of China*. Cambridge, Massachusetts, and London, England: The Belknap Press of Harvard University Press, 2011. Print. Online version, Accessed June 15, 2015.

http://www.gelora45.com/news/DengXiaoping_Transformati
onOfChina_Vogel.pdf

170-Warren, Kim and Robert Thurlby. "Understanding and Managing the Threat of Disruptive Events to the Critical National Infrastructure." **www.systemdyanmics.org/** January 23, 2012. Accessed July 6, 2015.
http://www.systemdynamics.org/conferences/2012/proceed/p apers/P1233.pdf

171-Waters, Brad. "10 Traits of Emotionally Resilient People." *Psychology Today*, May 21, 2013. Accessed July 11, 2015.
https://www.psychologytoday.com/blog/design-your-path/201305/10-traits-emotionally-resilient-people

172-Wilson, Tony. "The Dual Nature of Human Behavior." **Bath Royal Literary and Science Institution**, January 6, 2004. Accessed July 11, 2015.
http://www.brlsi.org/events-proceedings/proceedings/24929

173-Witherspoon, D.J., S. Wooding, A.R. Rogers, E. E. Marchani, W. S. Watkins, M.A. Batzer and I. B. Jorde. "Genetic Similarities Within and Between Human Populations." *Genetics,* May 2007 vol. 176 no. 1 351-359. Accessed June 9, 2015.
file:///C:/Users/Guest/Documents/Genetic%20Similarities%2 0Within%20and%20Between%20Human%20Populations.ht ml

174-Wolinsky, Howard. "Paths to acceptance. The advancement of scientific knowledge is an uphill struggle against 'accepted wisdom'." *EMBO Reports.* **US National Library of Medicine. National Institutes of Health**.
http://www.ncbi.nlm.nih.gov May 2008. Accessed July 15, 2015.
http://www.ncbi.nlm.nih.gov/pmc/articles/PMC2373380/

175-"World Population Growth." **World Bank**.
http://www.worldbank.org/ Accessed April 15, 2015.

http://www.worldbank.org/depweb/beyond/beyondco/beg_0
3.pdf

176-"World War II: Facts, information and articles about World War II, 1939-1945." **History Net**. Accessed June 16, 2015. http://www.historynet.com/world-war-ii

177-Zimmermann, Kim Ann. "Hurricane Katrina: Facts, Damage & Aftermath." **Live Science**. www.livescience.com/ August 20, 2012. Accessed July 11, 2015. http://www.livescience.com/22522-hurricane-katrina-facts.html

Bold facing is used for book titles and other sources to make reviewing references easier.

About the Author: Michael is an ex-alumnus of MIT's Guided Study Program in System Dynamics. He has worked in education as a teacher and high school principal, and traveled to Myanmar to create a System Dynamics group which provided training in Systems Thinking. He works as a Seller Support phone jockey at Amazon, writes at night, and lives in Costa Rica. He enjoys exploring, reading, writing and traveling.